T0184198

Drones for Good

How to Bring Sociotechnical Thinking into the Classroom

Synthesis Lectures on Engineers, Technology, and Society

Editor

Caroline Baillie, *University of San Diego*

The mission of this lecture series is to foster an understanding for engineers and scientists on the inclusive nature of their profession. The creation and proliferation of technologies needs to be inclusive as it has effects on all of humankind, regardless of national boundaries, socio-economic status, gender, race and ethnicity, or creed. The lectures combine expertise in sociology, political economics, philosophy of science, history, engineering, engineering education, participatory research, development studies, sustainability, psychotherapy, policy studies, and epistemology. The lectures are relevant to all engineers practicing in all parts of the world. Although written for practicing engineers and human resource trainers, engineering, science, and social science faculty in universities have reported that they find these publications an invaluable resource for students in the classroom and for further research. The goal of the series is to provide a platform for the publication of important and sometimes controversial lectures which will encourage discussion, reflection and further understanding.

The series editor regularly invites authors and encourages experts to recommend authors to write on a wide array of topics, focusing on the cause and effect relationships between engineers and technology, technologies and society, and of society on technology and engineers. Over the past decade and a half, topics have included, but are not limited to, the following general areas: History of Engineering, Politics and the Engineer, Economics, Social Issues and Ethics, Women in Engineering, Creativity and Innovation, Knowledge Networks, Styles of Organization, Environmental Issues, Appropriate Technology.

Drones for Good: How to Bring Sociotechnical Thinking into the Classroom
Gordon D. Hoople and Austin Choi-Fitzpatrick
2020

Engineering Ethics: Peace, Justice, and the Earth, Second Edition
George D. Catalano
August 2014

© Springer Nature Switzerland AG 2022
Reprint of original edition © Morgan & Claypool 2020

All rights reserved. No part of this publication may be reproduced, stored in a retrieval system, or transmitted in any form or by any means—electronic, mechanical, photocopy, recording, or any other except for brief quotations in printed reviews, without the prior permission of the publisher.

Drones for Good: How to Bring Sociotechnical Thinking into the Classroom
Gordon D. Hoople and Austin Choi-Fitzpatrick

ISBN: 978-3-031-00988-4 paperback
ISBN: 978-3-031-02116-9 ebook
ISBN: 978-3-031-00167-3 hardcover

DOI 10.1007/978-3-031-02116-9

A Publication in the Morgan & Claypool Publishers series
SYNTHESIS LECTURES ON ENGINEERS, TECHNOLOGY, AND SOCIETY
Lecture #24
Series Editor: Caroline Baillie, University of San Diego

Series ISSN   2690-0300 Print   2690-0327 Electronic

Drones for Good

How to Bring Sociotechnical Thinking into the Classroom

Gordon D. Hoople, University of San Diego
Austin Choi-Fitzpatrick, University of San Diego and University of Nottingham

SYNTHESIS LECTURES ON ENGINEERS, TECHNOLOGY, AND SOCIETY
#24

ABSTRACT

What in the world is a social scientist doing collaborating with an engineer, and an engineer with a sociologist, and together on a book about drones and sociotechnical thinking in the classroom? This book emerges from a frustration that disciplinary silos create few opportunities for students to engage with others beyond their chosen major. In this volume, Hoople and Choi-Fitzpatrick introduce a sociotechnical approach to truly interdisciplinary education around the exciting topic of drones. The text, geared primarily at university faculty, provides a hands-on approach for engaging students in challenging conversations at the intersection of technology and society. Choi-Fitzpatrick and Hoople provide a turnkey solution complete with detailed lesson plans, course assignments, and drone-based case studies. They present a modular framework, describing how faculty might adopt their approach for any number of technologies and class configurations.

KEYWORDS

team teaching, engineering education, pedagogy, drones, UAVs, design, peace technology

Contents

Advanced Praise

Drones for Good is brief, engaging, and written in an accessible style that does not get bogged down in academese. Hoople and Choi-Fitzpatrick are clearly enthusiastic about their project, informed about evidence-based pedagogies, and passionate about promoting sociotechnical thinking among their students.

> – Jon Leydens, Professor of Engineering Education Research, Division of Humanities, Arts, and Social Sciences, Colorado School of Mines

Candid and entertaining, *Drones for Good* is valuable for any academic motivated to integrate multiple perspectives into their course and empower students to address our world's greatest challenges.

> – Greg Rufilson, AAAS Science and Technology Policy Fellow at USAID

Hoople and Choi-Fitzpatrick guide us on a journey to where few academics dare venture: an authentic interdisciplinary learning community. *Drones for Good* removes the mystery surrounding effective interdisciplinary teaching, presenting the perfect blend of pedagogical design and practical implementation.

> – Tracy Kijewski-Correa, Leo E. and Patti Ruth Linbeck Collegiate Chair and Associate Professor, College of Engineering and Keough School of Global Affairs, University of Notre Dame

Acknowledgments

We are extremely grateful to a whole host of people, without whom this book would never have been written. It's wise to thank the money people first. We are grateful to work at a university where this kind of collaboration is possible. This collaboration was underwritten by an IUSE/PFE RED grant from the National Science Foundation (Award No. 1519453) and enthusiastically supported by our respective deans. Thanks to Caroline Baille, our editor at Morgan & Claypool, for her encouragement to write this book. Thanks to Beth Reddy for the unstinting research support that made this book possible.

At the Shiley-Marcos School of Engineering, we'd like to thank Dean Chell Roberts for introducing us to one another and then setting us loose on this adventure; the leaders of the RED grant for always pushing us to do better Susan Lord, Rick Olson, Michelle Camacho, Ming Huang, and Leonard Perry; administrators and staff for putting up with all of our wild requests Sam Burt, Steve Saxer, Elisa Lurkis, Choa Kang, Jocelyn Kuykendall, Michelle Sztupkay, Garry Frocklage, Lorena Silvas, and Paula Schmid; our student assistants who helped lighten the load, and the mood, along the way, especially Nick Cardoza and Khoa Vu.

At the Kroc School Dean Patricia Marquez has been consistent in her support of quirky and innovative approaches to Peace Studies. Austin's thinking about team science has been shaped by his ongoing collaboration with colleagues at the University of Nottingham's Rights Lab. We are in particular debt to a team of collaborators who helped build and refine a novel dataset on non-violent drone use, and who co-authored the published results, including Tautvydas Juskauskas, AlHakam Shaar, Dana Chavarria, Elizabeth Cychosz, John Paul Dingens, Michael Duffey, Katherine Koebel, Sirisack Siriphanh, Merlynn Yurika Tulen, Heath Watanabe, Mohammed Sabur, Lars Almquist, and John Holland.

Lastly, we'd like to thank everyone, especially our families and students, who put up with all of our crazy antics.

CHAPTER   1

Why Sociotechnical Thinking Matters

1.1   INTRODUCTION

What in the world is a social scientist doing collaborating with an engineer, and an engineer with a sociologist, and together on a book about drones and sociotechnical thinking in the classroom? This book emerges from a jointly held frustration that our respective disciplinary homes—engineering and public policy—have a lot to say about our specific domains of expertise but create fewer opportunities for our students to engage beyond a disciplinarily-bounded curriculum. As educators, we do our best to equip students for the moment when classroom knowledge inevitably competes against and clashes with the "real world."

The constraints of the classroom are not a new problem for academics. Hackathons, innovation competitions, and maker spaces are meant to foster new kinds of learning and collaboration. What these spaces and solutions rarely do, however, is engage actors beyond the disciplinary boundaries we ourselves as educators are trapped within. It is as if the world we equip engineers to enter is populated exclusively by engineers, and the same for those studying politics, economics, and society. Clearly this is not the fact. Engineers must work in the same ecosystem as marketing teams, lawyers, regulatory bodies, and, increasingly, ethicists. Students trained in social change tactics and strategies may end up in the employ of a non-governmental organization that has lawyers, computer programmers, and engineers on staff or on retainer.

As teachers will appreciate, but not every student fully grasps, the world is shot through with sociotechnical actors, factors, and phenomena. Cook stoves are created but never used, wells are dug but not maintained, and one laptop per child—a great slogan, if there ever was one—is also an effective, if unintentional, vector for pornography [1].

The lesson for engineers and others in the technical arts is that technical solutions enter into real social and political worlds. The lesson for those interested in politics and society is that social change efforts—to address inequality and climate change, for example—must engage complex realities that are both social and technical.

So how do these two communities—the technical arts and the social sciences—speak to one another? The former is relentlessly pragmatic and the latter is exhaustively reflective. The former might never look up from a workbench to ask how something will be used. The latter may never sit down to actually build something out of fear that it could fall into the wrong hands, or out of a lack of technical expertise in the first place.

Anyone possessing a passing familiarity with sociotechnical thinking will immediately recognize both the puzzle and the opportunity. Our objective in this book is to introduce one way to make this rather simple point come alive across disciplinary boundaries. It will not do to simply train engineers on ethics and social impact. Likewise, it will not do for students to pass through schools of public policy and social change without any opportunity to engage the raw mechanics that modulate much of the phenomena they hope to change.

In sum, a sociotechnical mindset is critical for students of social change as well as their colleagues in the technical arts. We wrote this book out of a conviction that while our colleagues may have every intention to teach this material, there are hardly any practical resources to help transform intent into action. We build on the work of our colleagues who care deeply about engineering and society—particularly Caroline Baillie, Donna Riley, Juan Lucena, and Jon Leydens—who have established the foundation upon which this text is built [2–4]. We hope this book provides tangible and practical next steps for faculty interested in bringing sociotechnical thinking to their classroom.

1.2   WHY DRONES?

In the pages that follow we develop a modular pedagogical approach that can be deployed *independent of the technology in question*. In other words, this is not strictly a book about a class on drones. It is instead a framework for engaging the complex interplay of social and technological factors in the classroom. The book's focus is the sociotechnical, rather than any particular technology.

For this exercise, we did choose a particular technology, and not an unimportant or uninteresting one at that: drones. The first reason we selected drones is that we both were already interested in the technology, both out of professional curiosity, but also out of a geek appreciation for toys, rockets, flight, and such. Austin was in the midst of a book on the social and political impact of drones, satellites, balloons, and kites and Gordon was ruminating over his time spent in aerospace consulting, where he worked on all manner of military drones. At a superficial level, we were curious about this new platform and certainly wanted to learn more.

The second, and more substantive, reason we chose drones as the class' project modality is that they truly embody and exemplify the interlinked puzzle technology and society represent. Technological innovation in the mobile telephony sector made lightweight drones possible, adding a new aspect to public and private airspace, and raising a host of legal and ethical issues. Drone technology, we decided, is intellectually stimulating, publicly disruptive, and personally engaging. We imagine many readers will agree with us, but are confident our approach to sociotechnical thinking can be taught at some other intersection of technology and society. How much personal data should mobile apps gather? What does responsible mining look like? What are the tradeoffs involved in programming self-driving cars? In what situations might gene editing be appropriate? Are robotic weapons systems ethical? Those puzzles require the kind of sociotechnical thinking we believe this book helps foster, and that we are confident our class modality can support. In other words, we chose to teach this class using drones, but expect others could do just as well with another topic or technology.

1.3   WHAT ARE DRONES?

We use the term "drone" throughout this book, even though there are much better technical terms on offer. Industry actors prefer terms like Unmanned Aerial Systems (UAS) or Unmanned Aerial Vehicles (UAVs). Others prefer configuration-specific terms, including "quadcopter" or "octocopter." The U.S. Military, eager to avoid the impression that "unmanned" devices are "unpiloted," advocates for the term Remotely Piloted Aircraft Systems (RPAS) since it makes explicit the fact that there is a human in the loop (HITL), a major requirement for weapons systems under international law. Austin prefers RPAS because it implies a system or platform (rather than just "vehicle") and because it avoids implying the pilot's gender. In the final analysis, however, we have found the term "drone" to serve as an excellent—and widely recognized—shorthand for these terms. Overall, we consider these terms to be largely interchangeable, except in those cases where we specify that a particular drone build was a quadcopter—it is then that we hope the reader recognizes that we are referring to a UAV/UAS/RPAS/Drone that has *four* propellers.

1.4   ORIGIN STORIES

First, a bit about us. We feel that introducing ourselves so early in a book is a touch self-indulgent. But we believe that recognizing and articulating one's own position and points of reference—one's positionality—is key to being a successful educator. Furthermore, we hope this context illustrates a larger point, which is that a project like this one requires unusual and diverse resources, as well as committed fellow travelers.

1.5   ABOUT US

Gordon is an assistant professor and a founding faculty member of the Integrated Engineering department at the University of San Diego's Shiley-Marcos School of Engineering. With an undergraduate degree in engineering and a Ph.D. in Mechanical Engineering, his experience in the field cuts across disciplines. During his Ph.D. he earned a minor in Science and Technology Studies, often finding himself to be the only engineer in classrooms filled with sociologists, anthropologists, and lawyers. His NSF-funded research focuses on engineering education where he is developing innovative pedagogies aimed at training students to understand the complex ways in which engineering impacts society [5, 6].

His interest in drones began with his work in industry. After graduating from college, he worked as an aerospace consultant performing modal analysis on a wide range of drones including the General Atomics Predator and the Boeing Phantom Eye. This first-hand experience forced him to consider tough questions about these and other defense technologies, eventually leading him to leave the aerospace industry and pursue a career in academia.

Austin is an associate professor at the Kroc School of Peace Studies at the University of San Diego, a concurrent associate professor at the University of Nottingham's Rights Lab and School of Sociology and Social Policy, and the co-founder of The Good Drone Lab. Trained as a sociologist, Austin has a background in human rights and social movements, and his interest in drones stems from an experimental human rights lab that he conducted with his graduate students while a professor at the School of Public Policy at Central European University.

With student colleagues at The Good Drone Lab Austin has argued for an ethical approach to the use of drones [7], estimated the prevalence in certain patterns of drone usage [8], developed and implemented a methodology for estimating crowd size [9], and written a book about how drones and other geospatial affordances democratize surveillance [10]. Austin was attracted to this collaboration with Gordon by an enduring sense that students need practical skills, combined with a love for tinkering and the arts. He's pretty convinced that the integration of the arts and sciences will help our research—and our research institutions—to better fulfill core objectives.

It should be clear that we have each had a foot firmly planted in our own disciplinary world, while also exploring the implications and opportunities our interests and research findings have for other spaces. That's why we co-direct The Good Drone Lab, an initiative that houses these efforts. We share a broad curiosity, rather than a keen focus on one particular puzzle—in other words, drones and sociotechnical education are one example of collaboration. We can imagine projects in other areas where our interests overlap (for example, we have started a collective called *Art Builds* with colleagues also interested in the intersection of the arts, humanities, and sciences) [134].

1.6   HOW WE GOT STARTED

In the fall of 2016 our colleagues received a National Science Foundation grant for Revolutionizing Engineering Departments [11]. They in turn asked the faculty for internal proposals and we pitched a class that would bring our respective students together. The project we pitched, implemented, and describe in this volume was a stretch for each of us—a bit weird for the sociologist and a bit weird for the engineer—but was fun and refreshing for us both. Our collaboration was organic, rather than imposed. Our deans created a space to find fellow travelers across campus, rather than forcing us to work together. In sum, funding from this grant freed us to create a class that identified opportunities for improvement within both of our educational ecosystems and freed us to experiment with various approaches that linked and blended the social and technical. Running in the background, we should emphasize, were our simultaneous commitments to doing work at the edges of our own—and trespassing a little into others'— disciplines.

This funding also provided valuable research assistance from postdocs and undergraduates. Across multiple versions of this course we collected quantitative and qualitative data in the form of focus group discussions, pre- and post-test assessments, and the analysis of student work. This all helped us to build the robust evidentiary base the observations in this book are drawn from. Over the past three years this collaboration has resulted in two classes, one workshop, a half-dozen co-authored publications, and this volume. Here we draw liberally on that data, subsequent findings, and scholarship we authored together and with others [8, 9, 12–16].

1.7   DEVELOPING A SOCIOTECHNICAL MINDSET

What's frustrated each of us in our respective fields is a broad disinterest in the kinds of ecumenicalism that makes science fresh and life interesting. This volume is directly informed by science and technology studies, whose focal premise is the importance of the interaction effects of scientific discovery and technological innovation on one hand, and culture, politics, and society on the other. This scholarly field is young and dynamic, and a glance at the front page of any newspaper or newsportal point to the complex interplay of technology and society. Is Wikileaks and Russian interference in the United States' 2016 election a story about technology, a story about politics, a story about society, or a stunning admixture of all three. Scholarship on Science, Technology, and Society (STS) has placed its money on the admixture, and we agree.

Neither academic consensus nor acrimonious debate automatically makes its way into the classroom. When we were first funded to develop a class that would draw our disparate students together around our shared interests, we looked to the literature. We found nothing on how to teach a hands-on class that reflects these core STS commitments. The book you are reading is the one we wish we'd been handed three years ago. We'd have found things to disagree with (as we anticipate you will as well), and plenty of places to customize and modify (as we hope you do), but we'd have used the book as a jumping off point for an in-class exploration of the interplay between technology and society.

Our experience developing this class convinced us of the importance—and accessibility—of a guide to holding both the technical and the social in one's mind simultaneously. This book thus advances a practical approach to developing a sociotechnical mindset. By *sociotechnical mindset* we simply mean the ability to identify and address issues with an understanding of the complex ways in which the social and technical aspects of these issues are interconnected. We argue that approaching things from a sociotechnical perspective is pivotal to addressing a number of challenges to our respective fields, including poverty, inequality, environmental degradation, climate change, urbanization, health, and so forth.

From our collaboration we have concluded that a sociotechnical mindset is also critical to the kind of educational outcomes we hope to achieve in the classes we teach (we are reluctant to use the term "classroom" as we feel most good learning happens beyond the room's edge). Such a mindset, however, is not something one can simply hand over to students. It must instead be arrived at through genuine collaboration between contrasting disciplinary fields—in other words, through an integrated approach.

We are, as one might imagine, not the first to make this argument. In 1955, the seminal Grinter Report called for a "continuing, concentrated effort to strengthen and integrate work in the humanistic and social sciences into engineering programs [17]." Fast forwarding to the present suggests that engineering education is still organized around "well-structured problems" amenable to resolution in the classroom rather than the "ill-structured" problems requiring creativity in the workplace [18]. Similarly, an Australian review of engineering practice determined that experienced engineers often find "the real intellectual challenges in engineering involve people and technical issues *simultaneously*" (emphasis added) [19].

This challenge is not unique to Engineering. Graduate programs focused on public policy, international affairs, and social change have traditionally focused on providing a theoretical and topical orientation in the field. The curriculum is complemented by skill-based training in three key areas: communication, operations, and key techniques. As a result, these schools provide skills training on marketing and media, management and leadership, and research methods. These training efforts are essentially limited to the tools available in the business and social science sectors, doing little to leverage complementarities with experts in the technical arts, including architecture, urban planning, health, medicine, water and sanitation, information technology, and the environmental sciences.

The social and the technical cannot be dealt with in different departments at opposite ends of campus. Instead students must be challenged to wrestle with complex, multi-dimensional problems within the context of a single integrated experience. Doing so will benefit engineering students, who will be provided an opportunity to better understand the world their work enters and impacts. It will also be beneficial to students at a school of Peace Studies. A focus on peace and justice emphasizes reducing violence and expanding dignity and rights. The approach we advocate here gives them an opportunity to think and talk about the technical arts styles and systems that are often treated as black magic—at worst—and a black box, at best, in curriculum focused on peace and justice.

1.8   OVERVIEW OF THE BOOK

This is not a book of sociotechnical theory. Plenty of excellent work exists, including Kaptelinin and Nardi's *Acting With Technology*, Bijker, Hughes, and Pinch's *The Social Construction of Technological Systems*, Latour's *Reassembling the Social*, Winner's *The Whale and the Reactor*, or, from a sociological lens, McDonell's excellent *Best Laid Plans* [20–24]. This book was instead designed for faculty at work in the classroom. We have taken every opportunity to introduce clearly explained concepts and easily implemented techniques that can be used by educators planning to emulate our approach in piece or wholesale. As a result, this document includes all the materials necessary to teach one's own version of our class. We also provide focused overviews of our assumptions, resources, and approaches. If we have been successful, then one should be able to easily identify what this book is doing and which bits of it are appropriate for one's purposes.

1.8.1   WHAT'S INSIDE

In the next chapters we focus on the practical details of how we organized and implemented a class that helps students to think sociotechnically. While we describe a course based around drones, we introduce a framework that we believe will allow this particular technology to be replaced with a topic of one's choice. We provide important contextual details about the class' origins as well as key aspects that improved the likelihood of success. Our course framework consists of three major elements:

1. *Thinking Sociotechnically:* A multi-week set of discussions and activities designed to help students see the importance of sociotechnical thinking. This section of the course looks similar to many social science courses (i.e., focused on texts and discussions).

2. *Technical Build:* A multi-week activity in which the students engage a technical task (in our class, building a drone). This section of the course is modeled on engineering design courses (i.e., building things in teams).

3. *Semester Long Project:* Linking the entire semester together is a project in which the students must approach a problem from a sociotechnical perspective. In our case, students were asked to design a drone that has a positive impact on a particular social issue.

In Chapter 2, we introduce our respective classroom philosophies and examine the value of a sociotechnical classroom. We then situate our work in contemporary scholarship on teaching, especially on the importance of moving away from the "sage on the stage" model of lecturing that remains prevalent in both of our fields, despite manifest evidence that it is inferior to more hands-on approaches [25]. This overview includes a discussion of our grading techniques, never a small matter when working across disciplinary conventions and with an eye toward motivating students.

In Chapters 3 and 4, we pivot to the brass tacks of the class, operationalizing some of the concepts introduced in Chapter 2, and providing a user's guide to the appendices. In particular, we discuss the ways we introduced students to "thinking sociotechnically." This process includes particular activities in disciplinarily heterogeneous teams, the creation of teams for a multi-week technical drone build, the work these teams did on semester-long projects for deploying their technical build, and an overview of best practices. We also provide details about the various build configurations we have experimented with as we worked to find the best fit for a 10–12-week class. We introduce the "cross-cutting themes" that allow faculty to leverage the class to deliver additional content on other topics. Finally, we discuss how we threaded entrepreneurship throughout our course, while also suggesting how other themes might be included, such as empathy (a previous focus) or social justice (a future goal). We conclude with some thoughts on how to choose one's own cross-cutting theme, suggesting some topics make better candidates than others.

In Chapter 5, we assess what makes drones a useful technology for this class, while also emphasizing the sociotechnical puzzles they raise. The current moment, we suggest, is an "unsettled time" for drones, as social, political, and regulatory norms have yet to stabilize in response to their emergence and spread. We propose a series of ethical guidelines that may focus classroom discussion around the use of commercial and hobbyist drones, as well as the deployment of the students' own platforms. We explore the implications of the current "unsettled" moment, especially as it relates to technological and social uncertainty about what final form drone technology will take and how social systems, like laws and norms, will respond. Chapter 6 is dedicated to a series of case studies that introduce real use cases and provide a number of questions for additional discussion. While most of this book is geared exclusively toward faculty, we wrote both of these chapters with the intention they could be assigned to students as reading.

In Chapter 7, we review the book's findings and recommendations. We also address a lurking issue—this class is premised on the importance of multi-disciplinary configurations in the classroom, but it is engineering that has captured the bulk of our attention throughout this volume. In this concluding chapter we suggest some of the benefits our approach brings to other disciplines, especially in the humanities and social sciences.

1.9   HOW TO USE THIS BOOK

We have designed this book to be useful. This means that it can be read linearly, as it appears in the table of contents and is outlined above. We know this is rarely how busy academics approach their work, as Austin is often scrambling to find key resources at the last minute (Gordon's usually a lot better prepared). That's why we've designed a handy *if-then* algorithm to help focus attention to the book's possible configurations.

If you want to teach this whole class from scratch, then:

- read this book,

- assign Chapters 5 and 6 as reading,

- use our syllabus, and

- review the Lesson Plan in Appendix 2 and follow as desired.

If you want to borrow our sociotechnical approach for inclusion as a module in one's own class, then:

- read Chapters 1, 2, and 3 and

- review our Syllabus and Lesson Plan for useful assignments and exercises.

If you want to use a few classroom exercises in one's class, then:

- review the Lesson Plan in Appendix 2 and

- review Chapter 3 to see how we used the exercise in question.

If you want to take a sociotechnical approach to design, then:

- read Chapter 4 and

- review the relevant activities in Appendix 2: Lesson Plan.

If you are an engineer who wants to integrate non-engineers into one's class, then:

- read Chapters 1, 2, and 3 and

- invite a colleague from across campus for coffee.

There are other configurations, but these should help you get started. In creating this book and these classes together—one engineer and one social scientist—we hope to have demonstrated the same process we invite our students into. We broke lots of silly rules in the process of creating these classes and this curriculum, and anticipate use of these resources will result in similar adventures.

CHAPTER 2

A Sociotechnical Education

2.1 INTRODUCTION

Assuming one is on board with our overall objective, how exactly does one go about creating a sociotechnical classroom? In this chapter we will detail our practical approach to this class. While this book leverages one particularly high-profile technology (drones), we have designed this project and written this text with the intention that anyone can apply our approach to most topical interests that lie at the social-technical intersection.

Fostering sociotechnical dialogue requires a diversity of viewpoints. We fostered this diversity by bringing together students from two disparate degree programs—engineering and peace. That composition itself sent an important signal that the class was different from others students may have attended, and challenged them to engage the material and their classmates in new ways. In both personal communications and anonymous evaluations, students reiterate a value for the perspectives of peers from other disciplines. While we had the benefit of bringing together a diverse student group, it is certainly possible to do a version of this class with all engineering students. What is most important is that the class feels in some way different to the students than their traditional disciplinary classes.

In this chapter we introduce our respective classroom philosophies and emphasize the value of a sociotechnical classroom. We also cover what we consider to be some of the best practices in education, as related to the major components in this class. In particular, we focus on the importance of: active learning that engages students in their own education; identifying clear and student-oriented learning objectives; and clarifying the relationship between critical pedagogies and our sociotechnical approach.

2.2 OUR CLASSROOM PHILOSOPHY

We share a commitment to a pedagogical approach that eschews the "sage on the stage" model. This commitment takes different forms in our respective spaces. At Central European University's School of Public Policy, Austin had converted a traditional Human Rights Advocacy class into a project-based effort that organized teamwork on two efforts: a traditional policy memo on minority rights and an untraditional methodology for estimating the size of mass gatherings (described in Chapter 5). Another class provided a semester-long opportunity for students to design and implement a research program focused on estimating the prevalence of purposeful nonviolent drone use (also described in Chapter 5). In both cases the classroom was converted into a project-based learning experience that led to the publication of research findings with student co-authors. Across both cases is a fundamental commitment to involving students in the real work of discovery and creation.

Working within the Engineering curriculum, Gordon also takes a hands-on approach. His goal is often to talk as little as possible during class—instead he gets out of the way so that students can learn things for themselves. He takes a holistic approach that simultaneously emphasizes analytical abilities and professional skills. This requires active pedagogies, such as group projects and reflection exercises, in all his courses. He works hard to create an inclusive learning environment across all of these modalities. This might be a good time to include a note about our students. The University of San Diego is a primarily undergraduate institution, although it offers a growing number of graduate degrees. The university is private, and while it attracts students from across the financial spectrum, it draws heavily on students coming from middle class and wealthier backgrounds. The University has 9,000 students in both undergraduate and graduate programs, and 38% of this population is students of color. These facts should be kept in mind.

2.3   THE VALUE OF A SOCIOTECHNICAL CLASSROOM

The value of a sociotechnical classroom has four components: Faculty, Students, Coursework, and "Other" phenomena that emerge in an idiosyncratic fashion. Our approach to our own involvement (Faculty) was fundamentally defined by our joint commitment to plan and implement every session together. With the exception of a handful of absences for conferences, we both participated in each of the classes and debriefed after every class. In other words, we planned, engaged, and debriefed together for virtually every class we offered. This approach is different from the turn-taking that sometimes accompanies team-taught classes. The result was an admixture of perspectives that came from our having hashed the reading out in our lesson-planning, debated over the class exercises in our specific class-prep, and captured key lessons after each implementation. This process helped us work together as colleagues, but it also ensured that the interdisciplinarity that came through in the class was the result of our own efforts, rather than a rote script we had inherited. This approach was important for us as a team, but it also provided a demonstration effect for our students. It also takes a lot of time.

Modeling interdisciplinarity for our students was crucial, as their engagement in the class can make or break the experience. We found students to be more receptive to our approach when it was clear we were enthusiastic and on the same page. We have also found that students are more receptive to faculty's ideas when it appears we are actually interested in, and believe in, our own ideas. This modeling effect cannot be overstated. As a result, students appeared increasingly comfortable as the semester went along, as they came to better understand our style and the class objectives, and as the class developed its own idiosyncratic dynamic. Clearly, a selection bias is at play, insofar as the class is attended by students who opted to take the class over reasonable alternatives. Here we should mention that in both of our Schools we spent not-inconsiderable time generating awareness of and interest in the class, such that we would have the opportunity to select from a waitlist of students who would best complement the class.

Perhaps most obviously, we selected readings that drew from each of our respective fields. This is perhaps the most explicitly interdisciplinary moment in the class, as engineers were asked to read key texts on the impact technology has on society, and non-engineers were asked to learn what it is that engineers actually do. The ensuing in-class conversations shed important light on key assumptions as well as opportunities for further engagement.

The final value of an interdisciplinary sociotechnical classroom lies in its emergent properties—in other words, the fact that *the whole is greater than the sum of its parts*. We find this conceptual error term useful in helping clarify the reason no two of our classes have been the same. In one year, we had a class comprised of students working for defense contractors as well as students preparing for the Peace Corps, and the teams worked together marvelously. In another class we had a team nearly implode because of tensions between a student committed to states' rights and another committed to universal human rights. While tensions are a rite of passage for any project-based teamwork, the admixture of multiple disciplines provides fresh opportunities for both delight and frustration.

At the broadest level, it may be that other forms of heterogeneity may suffice. It's even conceivable that such a course could work with only engineers—as long as an effort is made to bring engineers from different fields together on a problem that raises interesting challenges. We were fortunate that our class attracted engineers from three different areas of concentration—mechanical, electrical, and integrated. This too contributed to the teams' diversity. Each of these groups has a different definition of engineering, leading to interesting discussions and varied reactions to the assigned readings. Peace Studies students also came from a wide range of backgrounds. The Kroc School of Peace Studies offers master's degrees in Peace and Justice, Conflict Resolution, and Social Innovation, and attracts domestic and international students with degrees in the social sciences and humanities from universities around the world. This heterogeneity was a critical aspect of our class, and was the subject of much planning and discussion.

2.3.1 GETTING STARTED

If at all possible, we encourage others to explore team teaching with a colleague from a different discipline. Both of our Deans invested in this class' success, as evidenced by their willingness to underwrite two faculty members in a single course. This allowed us to both be physically present in the classroom at all times, making it possible to riff off of each other's comments and to demonstrate interdisciplinary behavior for our students. With the two of us present, we were also able to learn from each other's teaching practices. Austin is rather scattered, and only has one book about pedagogy on his shelf, entitled *Just In Time Teaching*, which he has never read. Gordon is a planner, owns three copies of a book called *Getting Things Done*, and countless books on pedagogical excellence. Needless to say, Austin learned a lot in this process. As a result of Gordon's expertise, we used a mix of activities—discussion, labs, project time, and lectures—that were common in each of our respective fields and are detailed in Chapters 3 and 4. While we leveraged our own areas of expertise, we were careful to always be active participants in the other's activities. We actively avoided falling into the common trap of team teaching in which individual faculty members are only responsible for their own content. While team teaching can be challenging, it is well worth the effort. For those interested in learning more, we recommend Plank's *Team Teaching Across the Disciplines* [26].

Scheduling was one major unforeseen challenge for this course. We discovered that the Engineering and Peace schools operate using very different student schedules. Engineering courses for undergrads tend to take place in the morning while Peace Studies courses for graduate students tend to be offered in the afternoon. We found that the only viable solution that did not conflict with other required classes was to hold the course in the evening. We choose to meet once a week for three hours—this gave us ample time to dive into both challenging sociotechnical discussions and complex engineering builds. A corollary to our scheduling challenge was that students also found it hard to meet outside of class—when the engineers were free the Peace Studies students were in class and vice versa. Therefore, providing time in class for students to work on their projects was pivotal—it was the only protected time when we could guarantee that students would all be available.

Another important consideration in the sociotechnical classroom is the learning space itself. The physical space where a class is taught has been shown to have a substantial impact on student learning [27–30]. Another benefit of teaching at night was that we had our choice of classrooms. We have had the benefit of teaching these classes in rooms designed for active learning and teamwork. The classroom we chose was set up with small tables designed for teams of four to six students. This setup was key to both facilitating discussions and designing and building drones. For many of our course activities we ask students to first engage at their tables with their teammates. Having students positioned at round tables made it easier to engage one another, though square desks have also worked (Figure 2.1). We also chose to hold courses in the Engineering building, as opposed to Peace. Indeed, one of our early activities was to have the engineering students lead the peace students on a tour of the engineering space—helping orient them to their surroundings.

Figure 2.1: One of the classrooms where we have taught Drones for Good.

2.4   BEST PRACTICES IN EDUCATION

Over the last three decades there have been tremendous advances in understanding how students learn. Scholars of learning science have done impressive research to uncover what factors motivate students, what practices do and do not work in the classroom, how students develop mastery, and what we can do as educators to promote self-directed learning [31]. Unfortunately, many college classrooms look the same today as they did when even the oldest administrators were in school. Most Ph.D. programs train students how to be researchers, not how to be teachers. In fact, both of us were dissuaded from pursuing additional training in teaching during our Ph.D. programs, and one of us was explicitly barred from doing so.

In this section we provide a short overview of the key pedagogical concepts upon which this course is based. For some readers this may be review—feel free to skip ahead. If, however, these are unfamiliar concepts, we hope this is your gateway an exciting new literature.

2.4.1   ACTIVE LEARNING

Active learning refers to classroom teaching techniques that, in some way, engage students *actively* in the learning process. The learning sciences research in this area has reached consensus: active learning leads to better student outcomes than do passive techniques like the lecture [32–35]. In his thorough review of active learning, Professor of Engineering Michael Prince argues for three categories of active learning: collaborative learning, cooperative learning, and problem-based learning [33]. *Collaborative learning* is any activity where students are learning together from each other. *Cooperative learning* involves any activities where students work together towards a common goal, but are assessed individually. *Problem-based learning* describes an approach where a problem is introduced to students and used to guide learning on a particular topic. We use all of these methods in our course, and discuss specific activities in the next chapter. In addition, the detailed lesson plans in Appendix 2 give examples of the activities we use to engage students in active learning.

At the heart of our commitment to active learning is the project-based learning (PBL) approach. Similar to problem-based learning, in PBL students are presented with an authentic, complex, open-ended problem and spend a substantial amount of time—up to an entire semester, as in our class—working on a solution [36, 37]. One particular feature of PBL is that work products should feel relevant and authentic—matching what students might be expected to produce by an employer after they leave college. PBL has been shown to provide a wealth of benefits to students [38–43]. By integrating problem- and project-based Learning approaches we designed a course that allows students from very different disciplinary backgrounds to learn from each other about their respective disciplines as they worked collaboratively.

2.4.2   LEARNING OBJECTIVES

As educators we often begin developing classes by thinking about the content we want to share with students. For example—if one is going to teach a course about drones, one could start with a list of topics that simply MUST be covered: battery power, national regulations, systems control theory, public opinion, and so forth. While content is important, however, what really matters is student learning. Learning objectives are a concrete set of goals that communicate to students what, exactly, they are expected to learn [44]. In addition to benefiting students, learning objectives are helpful for the instructor. They provide a useful benchmark when developing class materials, homework assignments, or exams. If a particular activity is not linked back to learning objectives, then that's a problem.

Ambrose et al. describe four key elements to consider when developing learning objectives [31]. Most importantly, learning objectives should be student-centered. When developing objectives, we like to think about what the students should be able to accomplish. This can be captured when writing objectives by using language such as "At the end of this course/assignment/class, students should be able to … ."

Second, learning objectives must focus on specific cognitive processes rather than complex ideas. Faculty, as experts, have often forgotten how complicated certain ideas can be for novice learners. In order to craft effective learning objectives complex tasks like problem solving must be broken down into component skills.

Third, learning objective should use action verbs that make the learning goals explicit to the students. Consider the passive verb *understand*—what, exactly, does it mean to "understand" something? Fortunately, verbs for learning objectives are well defined. In 1956, Benjamin Bloom developed a taxonomy of educational objectives that range in complexity from the simple—recall—to the challenging—*create*. These verbs can be easily found with a quick online search and provide a helpful framework when developing learning objectives [45].

Last, learning objectives should be measurable. It is critical that as educators we are able to assess whether or not our students have achieved our learning outcomes. To return to our previous example (and to highlight the complexity of passive verbs), how would one measure whether or not a student actually understands a concept? Clear objectives explicitly specify what the student should be able to do, for example: "Describe at least three different ways drones are used in civil society." Learning objectives are discussed more in Chapter 3 and provided in full in the Appendices.

2.4.3   TEAMING

There is a huge body of literature on fostering high functioning teams [46–50]. A full review of teaming is beyond the scope of this book, so instead we will simply highlight the teaming approach we have taken in this class. In our class we focus on building student teams that balance student interest and diversity. At the beginning of the semester we facilitate an exercise designed to generate a large number of ideas about how to use "drones for good." This process produces a pile of rough ideas, which students are then asked to refine. In the third week of the course we distribute a survey that asks students to rank order these ideas according to their own preferences. We use this information alongside student demographics to form teams of four students each. Student interests are considered, but are not the prime objective. The literature on team formation suggests that students have the best outcomes when women and minorities are not isolated on a team [51]. In other words, it is better to create one team with two women and one team with all men rather than to divide two women into separate teams, thereby isolating each of them. Forming teams in this way helps ensure that all student voices are present in spaces they can be heard, rather than that every team has all student voices.

We anticipate the research on ethnicity and gender might also apply to disciplines, in other words, that we might see similar problems if we were to isolate students from a particular degree program on a team. In response we create teams comprised of two engineers and two Peace Studies students. The final result is teams that strike a balance between interest, gender, ethnicity, and major. We should note that we do not tell the students what their shared interests are when we assigned the teams. Upon announcing the teams, we simply allow the students to continue their brainstorming and select a project that would work for all of them. We have never had students complain about being isolated because of their ethnicity or gender, nor have we ever had students complain that they were placed on a team that produced a project they were unhappy having contributed to. In the final analysis, teams are built around our commitment to best-practices in teaming, and student interests take a back seat to this consideration.

2.4.4   A BRIEF NOTE ON GRADING AND MOTIVATION

As anyone that has spent time in the classroom knows, motivation is key to student learning. But what, exactly, motivates students? Unfortunately, this complicated question has no straightforward answer. While grades play a role, students have a wide range of motivations—some intrinsic and some extrinsic. What motivates one student may be off-putting to another. Ambrose et al. describe a model of motivation that combines the subjective value students place on goals with their expectancy for success. They provide 18 actionable and research-based strategies to help motivate students [31]. Most pertinent to our course are a range of other commitments, including: connecting material to students' interests; providing authentic real-world tasks; demonstrating relevance to future professional lives; creating assignments with an appropriate level of challenge; providing targeted feedback; clearly articulating expectations; and giving students opportunity to reflect.

Grades are one example of an extrinsic motivator—one that is sometimes used by faculty as the only motivator. Gordon can't count the number of times he's heard students ask: "Will this be on the test?" (Austin can't remember the last time he gave a test.) Students are excellent problem solvers—we have trained them to find ways to get the best grades with the least amount of effort. Unfortunately, grades are not a particularly useful indicator of much beyond student's ability to get good grades. Research has demonstrated that grades are not well correlated with post-college success [52–54]. Furthermore, there is evidence that fear of low grades actually detracts from students' motivation [55, 56]. As Adam Grant, Professor of Management and Psychology at the University of Pennsylvania, argued in a *New York Times* editorial [57]:

> *If your goal is to graduate without a blemish on your transcript, you end up taking easier classes and staying within your comfort zone. If you're willing to tolerate the occasional B, you can learn to program in Python while struggling to decipher "Finnegans Wake." You gain experience coping with failures and setbacks, which builds resilience.*

While grades are a reality of the college experience, we minimize their importance in our classroom. At the start of the semester we tell students "We don't care what grade you get in this class, we care how much you learn." (See our syllabus in Appendix 1 for our full statement on grades.) We design our course in such a way that every student has the opportunity to earn an A and we make abundantly clear what our expectations are for A-level work through the use of detailed grading rubrics. One thing we struggled with when first offering our course was how to evaluate student's personal reflections. How, exactly, do we differentiate between an A-level reflection and a B?

We responded to this dilemma by adopting a grading technique known as "Definitional Grading" [58]. In this approach, student assignments are broken into two categories: graded work and pass-fail work. In order to earn a particular grade, students must achieve that grade level in both categories. *Graded work* is assessed in the same way as it would be in a standard course, using rubrics and an A–F-based grading system. *Pass-Fail work* is assessed simply by counting the number of assignments students successfully complete. For example, to earn an A one must pass at least 90% of the pass-fail work and earn A's on all graded assignments. Our specific approach can be found in the syllabus included as Appendix 1. We have found this approach helpful in signaling two things. First, it helps students see that we are not going to judge them on the specific content of their personal reflections. *As long as you put in a substantial effort*, we tell them, *you can earn a passing grade*. This helps students move away from simply telling us what they think we want to hear to actually developing and sharing their own opinions on the subject. Second, the students see that even though the work is not letter graded, it is still an important part of the course.

2.4.5   CRITICAL PEDAGOGIES

Critical pedagogies represent a heterogeneous group of approaches to teaching that seeks to empower culturally marginalized and disenfranchised students, acknowledges that traditional schooling works against the interests of those who are most vulnerable in society, and recognizes that knowledge is socially constructed within a historical context [59]. Critical pedagogies grounded in critical theory began to emerge in the 1970's. As our colleague Alex Mejia explains, "Traditional theory seeks to only understand or describe society, while critical theory seeks to critique and change society [60]." Much has been written on critical pedagogies and a wide variety of teaching approaches fall under this category [60–65].

Our sociotechnical approach borrows heavily from critical pedagogy in its emphasis on helping students develop their critical consciousness—the ability to critique cultural norms, values, and institutions that promote systemic social inequities [66]. Paulo Freire, Brazilian activist, educator, and one of the fathers of critical pedagogy, emphasized the importance of action based on reflection—a concept known as *concientização* (conscientization) [65]. Through dialogue and reflection, students, and their teachers, begin to understand the cultural forces that shape their lives and discover the ways in which they are empowered to create change. For students to approach problems from a sociotechnical perspective, it is imperative that they understand their own positionality—in what ways have they experienced power, privilege, and oppression—and to account for these factors in their own work.

2.5   CONCLUSION

Our respective classroom philosophies were important for how we found one another, how we chose to work together, and how our classes were designed, implemented, and iterated on. It is important to emphasize that across our respective teaching portfolios we have each done our best to incorporate many of the approaches specified in this chapter. Each of our independent commitments to these values make us natural allies in the creation and deployment of the class. This project was further facilitated by key support from our supervisors, as well as generous financial and in-kind support from both an external grant and our home institution. This is not to toot our own horn, but to instead be frank about the conditions under which this class was successfully deployed. We stand on the shoulders of very supportive administrators, as it were.

In this chapter we also set out to clarify our understanding of the relevant best practices in education, paying particular attention to how they impact our class design and implementation. The class is premised on the importance of student engagement in the education process. We work hard to ensure that each of our learning objectives is clear, action-oriented, student-focused, and auditable. Significant effort has also gone into creating content and a process that involves students in the process of clarifying goals, specifying tasks, creating systems of accountability, and staying on schedule.

CHAPTER 3

The Sociotechnical Classroom

3.1 INTRODUCTION

In the preceding chapter we focused on our pedagogical orientation and philosophy. In this chapter we move to the nuts and bolts of how we actually implement this class. In broad brushstrokes there are three major elements of our class. During the first few weeks we focus on helping the students to think sociotechnically. This portion of the class feels somewhat familiar to the non-engineers in attendance, since we ask the students to do readings and hold discussions in class. The interdisciplinary nature of the classroom, however, dramatically changes the tenor of these discussions. In small groups we ask students to reflect on their disciplinary identity, to think about the ways in which problems don't know disciplinary boundaries, and challenge them to think about the ways in which technology and society interact. Engaging the reading through these questions is uncharted territory for all of the students.

In the next portion of the class we shift out of reflection and into engagement, as students team to build a drone. At this point the students from engineering begin feeling more comfortable, but the interdisciplinary nature of their teams forces them to slow down and figure out how to explain many concepts they had taken for granted. At the end of the build we hold a friendly and un-graded flight competition. In the final phase of the class, student teams must develop, design, and pitch a pro-social use for a drone. They needn't build the drone, but they must develop a Minimum Viable Product (more on these soon) and present it as a part of their final pitch.

In the next two chapters we go step-by-step through what exactly happens across these three phases. In many ways this chapter will serve as a guide for Appendix 2, where were we provide a detailed lesson plan that covers what we do in every minute of every class. We encourage the reader to flip back and forth between these sections, looking in the main text for an overview and moving to the Appendix for the details of a particular activity that might be of particular interest.

Before we dive in, it seems useful to provide a bit of institutional context. The University of San Diego is a small, private, liberal arts, Catholic, predominantly undergraduate institution. With an emphasis on teaching, the university is philosophically committed to small classes—few rooms on campus can even hold more than 35 students. In our case we have capped enrollment to Drones for Good to just 24 students—6 teams of 4. While this is larger than the median class size at our institution, we think many of the approaches we describe could be scaled up for larger classrooms.

3.2   HELPING STUDENTS THINK SOCIOTECHNICALLY

The first five weeks of our course is dedicated to helping students develop their own sociotechnical toolkit. We begin the semester by asking students to reflect critically on their own disciplinary identity. For many students this is something they have never done before. What makes the exercise work is the fact that the classroom is composed of two groups of students who know very little about each other's discipline. Trying to explain concepts of engineering and Peace Studies across this disciplinary divide forces students to engage and understand what their own discipline is really about, recognizing that *what you see depends on where you stand*. After exploring their own positionality, we challenge students to: think about how social change happens; explore the idea of social innovation; discuss power, politics, and inequality; and discuss emerging technologies. We cover one of these topics per week. At the same time, we also get the students started on their major projects, discussed later in this chapter. Our learning outcomes are realistic. At the end of the semester students should be able to:

1. define sociotechnical dualism and explain to a peer why they should approach problem solving from a sociotechnical perspective, and

2. debate issues related to social change; social innovation; power, politics, and inequality; and emerging technologies.

We work toward these learning outcomes by offering lessons intended to help students see the way in which the world tends to categorize problems as either social or technical. We further help students to see why it is imperative to approach problems from an integrated approach. Our classroom activities themselves look similar to what might traditionally be found in humanities or social science courses—but take advantage of the two different disciplines within the classroom. The next sections explain our key pedagogical techniques for this segment of the course.

3.2.1   IN-CLASS ACTIVITIES

Inviting Discussion

Even with a small class of 24 students, it can often be hard to have class discussion in which everyone can participate. One approach we take is to get students started in small groups of four or five. We begin many of our classes asking students to discuss a few warm-up questions in these smaller groups. For example, in Week 2 we ask each group to summarize one of the previous week's readings and then report out to the class. The dynamic of moving between small and large groups helps encourage participation, especially from those who are comfortable speaking in the small group but are intimidated by the larger classroom setting.

Structured Debates

We have found structured debate to be a very powerful tool, particularly in an interdisciplinary classroom. We organize this activity by first asking for three volunteers to be judges and we then arbitrarily divide the remainder of the class in half, ensuring each team has both engineers and Peace Studies students. We instruct each team to argue a particular side of an issue. Each team is then given time to prepare their argument, and then we hold the debate. Each side is given time to present initial arguments as well as a rebuttal to the argument raised by the other side. We usually take a short break after the debate, giving the student judges times to confer and choose a winner. Care is taken to remind the judges to evaluate the debate based on the merits argued, not their personal belief on the topic. One particularly engaging topic for debate is whether technology benefits society. Surprisingly, it is often the engineering students that make the most compelling arguments for the ways in which technology has negatively impacted society.

Flip Chart Round-Robin

In this activity, before class begins, we choose six discussion prompts and write them at the top of separate pieces of flip chart paper. For example, one topic we have used is: *Who should help manage trade-offs in technology's risks and rewards, governments, businesses, activists, citizens, or consumers?* We then position these prompts at six different physical locations throughout the classroom. Once class has begun we give the groups five minutes to discuss the prompt at their table and capture their discussion on the flipchart. We then have each team rotate to the next topic, repeating the process until they have responded to every prompt. At each stage students are able to read what has previously been written on the topic, thereby jump-starting their debate. At the end of the activity teams return to their original station, synthesize the notes on their assigned topic, and report out to the class. We have found this to be a particularly high energy activity, perhaps in part because students are always on their feet and moving. It is an excellent choice for engaging students on a wide range of questions.

In the Public Eye

This is one of two activities where we divide students based on their disciplinary identity. The activity begins with all students from one of the disciplinarily homogeneous groups (e.g., all the engineering students) sitting in a circle in the middle of the room. This group is given a discussion prompt and instructed that everyone *inside the circle* must say something. Beyond this initial instruction, we do not intervene in the discussion. At the same time, the second group of students is instructed to encircle the first and pay close attention—in particular we direct members of the observing group to focus on the content of the discussion, but to also watch who speaks, note body language, and assess the overall dynamics of the discussion process. After the inner group has spent 5–10 minutes discussing the topic we conclude the round and ask the observing students to report what they saw. When this process has finished, the students switch roles and the new inside circle discusses a new topic.

What makes this activity particularly engaging is that we give a discussion prompt focused on engineering to the peace students and a peace topic to the engineering students. Being stuck on the outside and unable to speak about a topic where they have something to say is often a novel experience for the students. Perhaps the most important part of the activity is debriefing the process—we ask students what it felt like to watch and how it changed their behavior. We found this to be a particularly rich activity for discussions of power, politics, and inequality, since the structure of the activity reflects the themes we are talking about in class that week. For example, the question *What did it feel like to be powerless to comment on the discussion?* can open a rich conversation about how various forms of knowledge are privileged in our disciplines, on our campus, and in our societies.

Disciplinary Breakout

The disciplinary breakout is the only other activity where we divide students by discipline. We feel that it is important to dedicate some time for students to meet in discipline-specific groups. Our goal is to provide students the time to reflect, share, and, if necessary, decompress. Although we have not experienced significant tensions between students from our respective schools, planning at least one such reflection time over the course of the semester ensures that students have a release valve and that we have an opportunity to check in on the progress of their team-based efforts, as well as their individual experience of the class. Perhaps unsurprisingly, we have found the Peace Studies students are more actively engaged in this conversation. These conversations are held in the final third of a class period sometime mid-semester. We each facilitate the conversation with our own students, meaning the spaces are completely homogeneous.

3.2.2   OUT-OF-CLASS ASSIGNMENTS

For this portion of the class, we ask students to read several articles every week and write reflections about those articles. This requirement plays an important role in ensuring students engage with the readings. The engineers often struggle with the amount of reading in this class, so one of our earliest activities involves the peace students sharing reading strategies with their colleagues. This usually leads to fruitful discussions about things like reading for major themes rather than trying to comprehend every single word in article, a minor revelation to some of our engineering students. The major deliverable for this portion of the class is a three-five page essay (single spaced) that invites students to reflect on their discipline. We ask the students to examine their disciplinary lens and to identify three key themes from the class that have impacted their thinking (the full prompt is in the syllabus in Appendix 1).

3.3   BUILDING A DRONE

After several weeks spent primarily in the social sciences, the engineering students are quite excited to arrive at the build portion of the class. The main objective in this section is to give students some hands-on familiarity with drone technology. As we will soon discuss in more detail, the semester long project is aimed at designing a viable drone concept and producing a compelling pitch. To the students we stress that the build process is about gaining experience with this emerging technology and having fun. The build portion of the class culminates in a friendly flight competition where we challenge the students with an obstacle course. As a result, the learning objectives stipulate that after these lessons students should be able to:

1. assemble, test, and fly a quadcopter using off the shelf components and open source software;

2. explain to a friend the major ethical concerns related to drones; and

3. describe the key regulatory guidelines related to drones.

In the next sections we describe the various approaches we have taken to the drone build. Rapid technological advances have forced us to reconfigure our build every year that we have taught this class. We have found a number of kits made for the educational market, but they are all outside the price-point supported by our university's average lab fees.

Drawing on our educational experience, we have partnered with NewBeeDrone, a well-regarded manufacturer of indoor FPV drones, to develop a standardized kit that can be used to teach students about drones. The kit has been designed specifically for the education market, with the goal that drone builds can be seamlessly integrated by any instructor into the classroom. In addition to providing the hardware, we also provide step-by-step labs that help students develop a deeper understanding of this technology.

Before we continue to the build, an important word about safety is in order: drones can be dangerous. While many drones may look like toys, their propellers spin at thousands of RPM and can cause serious injury—especially to fingers, eyes, and hair. Drones should never be flown over 400 ft, out of one's line of sight, at night, or over people. Many densely populated areas, especially those with nearby airports, have airspace restrictions that prohibit flying drones of any kind outdoors. As this technology evolves, so too do FAA regulations. At the time of this writing any commercial use of drones, including academic research, requires a Remote Pilot Certificate under the FAA's Small UAS Rule (Part 107). Students at accredited educational institutions, however, are permitted to fly drones without a license under recreational guidelines [67]. To alleviate many of these concerns we recommend starting off by flying indoors with lightweight models (less than 100 g). Indoor flight mitigates nearly all of the substantial risks associated with outdoor flight and is not regulated in any way by the FAA. Indoor drones are still incredibly powerful and provide a great introduction to the technology. The kit we have developed with NewBeeDrone is designed exactly with this in mind. Before flying, please be sure to consult with online resources such as www. faa.gov and www.knowbeforeyoufly.org to confirm the most recent regulations regarding drone use.

3.3.1   THE BUILD

We have run three different versions of the drone build (Figure 3.1). Each has advantages and disadvantages. While the platforms have differed, the overall structure of the activity has remained the same. We typically begin the build by providing students with the components for the drone, but without any direction about what the pieces are or how to assemble them. This often causes some confusion and frustration, our intended result.

Figure 3.1: The three different versions of drones used in our course so far. From left to right: 270-mm quadcopter (custom built), Eachine E010 (commercial), and 90-mm microquad (custom built).

Most of the engineering students in our class are seniors, and will soon be practicing engineers. Since we want them to become stewards of their own learning, we push them to find their own resources, predominantly online, to answer some basic questions about drones. We provide these components towards the end of our three-hour class period and give the students about an hour to try and put the pieces of the puzzle together. As an added challenge we require each team to explain the purpose of the individual kit components. The twist is that the presentation must be given by the non-engineering members of the team. This requirement forces engineers and non-engineers to gain clarity internally (about who knows what, and what the pieces are), forces engineers to explain things to non-engineers, and forces non-engineers to explain basic engineering concepts to others.

At the start of the next class period we present on drone basics. This involves a half-hour presentation on key components, the basics of flight, and an overview of key safety issues. Before turning the students loose to start building their drone, we emphasize that the engineering students are to act as coaches rather than the primary builders. While we were initially concerned that the engineers would dominate the build, our experience has been the opposite. We have found the engineering students to be great teachers, taking pride in sharing their expertise in solving technical problems with their Peace Studies colleagues. Depending on the complexity of the drone, this build phase can take anywhere from a single class period to multiple weeks.

 Our most complicated version of the build was a quadcopter with a 270-mm custom designed wood and metal frame. We had teams of four work together to build these drones. The build process took roughly three class periods (about nine hours) and, due to the large-scale nature of the drone, required a field trip to fly at a local park. The drone used 4 brushless DC motors, 130-mm propellers, a 3S LiPo battery, and a CC3D flight controller. Students used the open-source software platform LibrePilot in order to adjust the settings on the drone [68]. While this drone was an excellent vehicle for learning, we do not recommend building something so large for students unfamiliar with drones. The primary reason is safety—while no one was injured, we had several incidents that were too close for comfort. In addition, the students were well aware of the risks associated with this drone and it made them shy away from flying. Lastly we found that, at least during the build, a team of four was too many—for many parts of the build only two people could realistically have their hands on the hardware and be actively involved.

Our simplest version of the build occurred during a compressed two-day workshop version of our course. In this case we selected an off-the-shelf nano quadcopter, the Eachines E010. We bought these ready-to-fly toy drones off-the-shelf and then disassembled them. Building these drones took roughly three hours and involved basic physical assembly: students simply needed to determine the right way to physically attach the various components, connect the propellers to the motors in the proper orientation (a non-trivial task), attach the battery, and start flying. The build did not require any programming, so no time was spent fine-tuning configurations on flight-control software. For this version of the build we put students in groups of two—this worked well as students could still learn from each other, but there was no reason to be disengaged. This was a successful build given the time constraints, however for a semester-long course we would recommend a more complex build. This is particularly important for engineering students as the simple plug and play assembly is often viewed as too simple and not "real engineering."

Our most recent version of the build was a mini quadcopter with a 90-mm carbon-fiber frame and 55-mm propellers. Using a 1S Lipo battery with brushed DC motors, this drone was primarily designed to be flown indoors. Assembling the hardware was slightly more complicated than the Eachine build, but what made this experience substantially different is that students connected the drone to a computer and programmed it using the open-source software Betaflight [69].

In addition, while we still have the students work together in teams, we provided individual kits so that every student could build their own drone. This gave each student a sense of ownership and also made it possible for them to take their drone home so they could continue practicing. Practice at home, and our on-campus flight competition, required students to tweak the platform's hardware and software. While some students were comfortable making these changes, others got frustrated and simply gave up. The major challenge we faced with this build was unreliable hardware. Several of the flight control boards we purchased were faulty and the motors and frame were fragile and broke after inevitable crash landings.

Each year, as we prepare to teach the course, we face the time-consuming process of sifting through hundreds of components to try and put together the optimal build for our students. The drone market is moving incredibly quickly, with new hardware and lower prices being introduced on the timescale of months, rather than years. More than once we have ordered components over the summer to prototype a drone only to find that when we were ready to place an order for the full class the hardware had been sold out and been replaced by something new. This steep learning curve is one of the reasons we have begun a collaboration with a drone startup to develop a standardized kit (there are now several similar kits available on the market). We recommend adopting one of these kits or finding local partners with drone expertise—be they hobbyists or educators—to help lessen the burden of adopting novel technologies.

3.3.2   FLIGHT COMPETITION

The flight competition is always one of the most enjoyable class sessions since it gives students a chance to relax and reap the rewards of their labor. We have held the competition both indoors and outside. We recommend the indoor flight competition for the safety reasons discussed previously. For those skeptical that indoor flying can be fun, we encourage you to simply search YouTube for *indoor fpv drone racing*. We make a point of having a friendly competition—students are not graded on their performance and are encouraged to simply enjoy themselves and compete for bragging rights.

We have found that setting up an obstacle course for the race substantially improves the students' experience, as does showing a few drone-racing videos to motivate the students to practice flying. One of the major learning outcomes from the flight competition is that students see how important pilots are to a drone's performance. Students often have the misconception that all drones are completely autonomous and are trivial to fly. While this autonomous flight is possible in certain well-controlled scenarios, trying to fly the drone they have built (with no GPS stabilization) through a hula-hoop goes a long way in convincing them just how challenging it can be to control these devices.

The flight competition also presents a nice segue to discuss the case studies, presented in detail in Chapter 6. One of the challenges we have observed is that students often struggle to identify the major ethical challenges associated with drones. We had thought with an emphasis on sociotechnical thinking in the first portion of the course, as well as with the peace students' background deconstructing technology, questions around the ethics of drones would naturally emerge during the build portion of the class. To our surprise, this was not the case. Therefore, as a new addition to the class, we intend to incorporate the case studies in Chapters 5 and 6 alongside the technical build and flight competition.

Figure 3.2: One of our favorite flight competitions happened to land on Halloween, adding to the excitement of an already fun evening.

3.3.3   ASSIGNMENTS

During this phase of the course, in-class and out-of-class assignments blend together. Depending on the complexity of the build one chooses, it is likely that only a few students will finish their build in class. Our goal is always to set the students up for success so that they can continue to build on their own at home. This is one of the reasons we have moved to an individual kit model, in which each student has their own drone. When building the drone required a team effort, it was particularly hard for students to find the time to meet outside of class. In addition to building a drone, we also asked the students to practice flying at home, since the most important thing for becoming a competent pilot is simply logging airtime. Lastly, we recommend incorporating case study reading alongside the drone build. This, in addition to the work students are doing on their drone for good project (discussed next), keeps them plenty busy.

3.4   CONCLUSION

This chapter operationalizes how we put our sociotechnical approach into practice in the classroom. We argue that *the whole is greater than the sum of its parts*, a fact which leads us to believe that unique benefits to this approach emerge on a case-by-case basis. What makes our space unique (strengths as well as weaknesses included) might lead to emergent phenomenon that are different from what another unique space might produce elsewhere. That, in many ways, is the value of a sociotechnical approach: it provides a controlled way to bring a bit of real-world complexity into the pedagogical frame.

In summary, we organize the class to engage students in a range of activities. These activities rely heavily on teams, as early discussions and subsequent drone builds happen in small heterogeneous groups. We less frequently rely on homogeneous clusters of students within their disciplinary groups, though in both of the times we have chosen this approach it yielded fruit, whether through observation (*In the Public Eye*) or in isolation (*Disciplinary Breakout*). Rather than jumping directly into classroom debates, we start with smaller conversations nested within the teams. We believe this gives students an opportunity to prepare for structured debates as well as unstructured conversations. In a similar fashion, we gave students the opportunity to discuss their reading and share reading tips, in these smaller groups. We think this approach increases the students' level of comfort when it comes to discussing their ideas in front of the class.

The assignments themselves provide a significant range of motion for the instructor. We have toggled through a number of drone platforms, each of which presented different tradeoffs between learning outcomes and costs (e.g., safety, time, frustration). This is clearly an area where others may choose different platforms, as we ourselves are through our own industry partnership. Over time we have refined our flight competition and feel this is an incredibly fun way to spend half a class. We imagine future iterations on other campuses could orient the entire class around this flight competition, for example, or do away with it altogether. Likewise, our decision to have students work in their teams to build each person's individual platform might be swapped out with the entire class building one single platform, or any other permutation one might imagine. Each of these configurations, we believe, facilitates sociotechnical thinking across disciplinary divides. Rather than being a blueprint, then, this chapter is meant as an example, illustration, and jumping-off point for tweaking and tailoring on the implementation end. The next chapter is focused entirely on the so what question that sociotechnical thinking begs. Students can now think critically and build effectively, but to what end? The next chapter presents our answer to this question.

CHAPTER 4

Designing a Drone for Good

Quadcopter by Randall Munroe (More at xkcd.com)

4.1 INTRODUCTION

Running through our entire semester is a project where the students must approach a problem from a sociotechnical perspective. We have chosen a particular project: imagining, designing, and prototyping a pro-social use for drone technology. By *pro-social* we simply mean uses that, at a bare minimum, comply with the Hippocratic Oath, which stipulates that's something must *do no harm*. We emphasize that we mean harm to anything—people or planet—as the environment as this is an area students often neglect. We are not ethicists—though we suggest some ways for debating drone use. Rather, we are writing from a sense that popular attention has focused on (1) military drone use and the ethics of warfare and conflict and on (2) commercial drone use and questions of safety and privacy. Those conversations are important, but we want to spark a larger conversation about how the non-profit organizations, universities, news agencies, and other public actors who comprise *civil society* can and should be using new technology. A larger version of this argument can be found in Austin's book *The Good Drone*, in which he argues that geospatial affordances like satellites, kites, balloons, and drones are fundamentally democratizing surveillance, and equipping civil society groups and individuals to do things that only big businesses and governments had been able to do.

How should we design drones that do no harm? That is the challenge we level to the students. In particular, we apply this question to the early stages of the design process—problem identification, ideation, and prototyping. Unlike many design courses, however, the final product for our class was explicitly *not* a piece of hardware. Instead, we challenge the students to plan a use, then prepare a compelling pitch for why someone should support their idea. This requires the students to fully flesh out the key elements of their concept—not only must they understand whether it is technically feasible, but they must also understand the social implications of their proposed technology.

We introduce this project on the very first day of class and every week we challenge students to make progress on some part of their design. While we have organized this class around the development of a sociotechnical mindset, we are also committed to students developing additional skill sets. As a result, we have built key skill modules into the class. These modules accommodate whatever tasks the educator has chosen. In this way the class is designed to accommodate a *cross-cutting theme*.

In this chapter we present the version of our course that focused on the cross-cutting theme of *entrepreneurship*. We developed skill modules that challenged students to think about the financial side of their design, exploring, for example, *how exactly will you find funding to create this drone for good*? The majority of this chapter describes our classroom approach and is complemented by the detailed course lesson plans included in the Appendix. At the end of the chapter we explore what other cross-cutting themes might be integrated into this course.

One of the best practices for problem-based learning is to make the experience seem as authentic as possible [70]. In a systematic review of authentic design experiences in engineering education, Strobel et al. argue that authentic-seeming problems are puzzles whose "primary purpose and source of existence is not to teach or provide a learning situation" but is rather an actual "need, a practice, a task, a quest and a thirst existing in a context outside of schooling and educational purposes" [70]. One way we have done this is to design our course to align with the submission requirements from the Fowler Global Social Innovation Challenge (GSIC) [71]. The GSIC is a social venture pitch competition that recognizes, resources, and rewards student-led social ventures focused on sustainable change. Started at USD, this competition has now involved 900 teams from 35 universities and has awarded over $350,000. This pitch competition is open to anyone from a degree granting university and we encourage anyone to consider participating. While we do not require our students to submit to this competition, many of the deliverables for our class are intentionally matched on the submission requirements for this competition.

One important note: in our class we invert the typical engineering design process. We present students with a particular solution—drones—and ask them to find a problem that can be solved with this technology. This is the equivalent of the old adage "when you have a hammer everything looks like a nail." We emphasize to students that best practices in design involve looking for problems and then exploring a broad solution space that considers multiple technologies. We make clear that we are instead flipping the script in this class because we want them to gain experience thinking about how drones in particular interface with society. It's not best practice in product design, but it helps the class get down to the brass tacks of heterogeneous teamwork in short order.

4.2   CLASS ACTIVITIES AND STUDENT ASSIGNMENTS

In Chapter 3 we distinguished between "in class" activities and "out of class" assignments. This distinction is not particularly meaningful for the project itself. We introduce all of the project activities in class and give students some class time to work on them. To finish the assignment, however, students must complete work outside of class. We challenged students to find ways to effectively manage their own time, emphasizing that what really matters is their output, rather than the amount of time spent or simply logging long hours. After these lessons, students should be able to:

1. design a drone that can have a positive impact on society (with a critical understanding of the ways drones can negatively impact society);

2. collaborate on an interdisciplinary team to solve problems and communicate ideas in a compelling way; and

3. acquire new knowledge to address problems outside of their comfort zone.

4.2.1   WHAT ARE DRONES ANYWAY?

On the very first day of class, our very first activity about drones challenges students to answer three questions: (1) What is a drone? (2) What are your concerns with drones? (3) How have you seen drones used? We start this activity by asking everyone to jot their individual answers, thus providing each student some time to think before we ask them to share. We then challenge the students to introduce themselves to a neighbor from a different program and discuss their answers to the question. Next, we lead a larger class discussion aimed at a identifying some larger trends and areas of confusion. This conversation usually allows us to raise a few interesting points.

In response to the first question, we find there is a lot of confusion over what counts as a "drone." We highlight the way in which the word drone is used quite differently by different groups. For example, while the media talks about drones, the U.S. Military prefers to speak of Remotely Piloted Aircraft Systems (RPAS). This question also helps get at the range in scale from large military-style aircraft such as the General Atomics Predator to commercial quadcopter platforms such as the DJI Phantom. We explain to students that we have chosen the term drone in part for its ambiguity. As they think about designing a *drone for good*, we emphasize the range of platforms and configurations this phrase invites.

Probing student's concerns with drones immediately raises many of the questions students will face throughout the semester, including: privacy, security, regulation, and ethics. Asking about how students have seen drones used gets students to start thinking about how they might want to use a drone to have a positive impact on society.

4.2.2   DRONE MINI-PITCHES AND TEAM FORMATION

Students' first homework assignment is to generate ideas for how drones could be used for the greater good. We stipulate that this should be an idea they want to work on for the rest of the semester and ask them to create a compelling pitch to convince their classmates to join a team focused on their proposal. Students are limited to 30 sec and the video can only show them talking, meaning no graphics or special effects are permitted. Videos are collected on the platform Flipgrid, but any video platform will do [72].

During the second class we watch all of these video pitches as a class. This activity serves two primary functions. First, it is helpful in focusing attention on a handful of uses. As we watch the videos several themes for using drones emerge. We use these themes to survey students about their preferences, and to inform our team-building efforts. Using this survey, we then form teams, balancing student interest with best teaming practices. Notably, we do not require the teams to work on their assigned theme—we simply used it as a grouping tool to bring together like-minded individuals. More often than not, teams end up working on a totally different project than originally pitched in their videos. This process and rationale are described in greater detail in Chapter 2.

This activity also gets students thinking about how to pitch ideas—a key goal for class in the Entrepreneurship configuration. We intentionally provide very little information about what makes a compelling pitch. As students watch the videos, they immediately begin to see which strategies work well and which fall flat. After watching the videos, we lead the class in a discussion about what they found to be the most effective pitch strategies. This often provides a good opportunity to explore disciplinary differences. We usually find some consistent trends in the styles adopted by the two groups of students. For example, the Peace Studies students often have compelling motivation for their project, but fall short on the details of how a drone might be used. The engineers have the opposite problem: they paint a picture of an amazing technology, but have trouble connecting their idea to some larger motivation.

4.2.3   IDEATION

Successful project topic ideation is one of the most critical activities of the semester: it lays the foundation for the bulk of the team's work. Much has been written on the topic of ideation and there are many activities designed to help teams generate ideas. Our approach draws heavily on work in design thinking, as championed by Tim Brown of IDEO and Stanford's d-school [73, 74] and draws on ideation resources from the KEEN Foundation [75, 76].

We start with the warm up activity "Yes, But/Yes, And." In this activity one student is selected to be the idea generator (this person can be easily assigned by nominating person with the nearest next birthday). This person is asked to generate ideas related to a low stakes topic such as planning a birthday party. During the first part of the activity the rest of the group responds to all of the ideas with the phrase "Yes, But…." and provides a reason why the particular idea will fail.

After 2–3 min the team usually gets the point—it isn't very fun to constantly have one's ideas shot down. We then repeat this activity. The topic and idea generator remain the same, but the remainder of the team must now respond with "yes and..." statements that support and building on the proposed ideas. The most important part of this activity is actually the discussion. We ask the idea generators to share how it felt in the first version of the exercise as compared to the second version. As one might expect, students report it is much easier to come up with ideas when everyone is saying "Yes, And..."

We use this exercise as a jumping-off point to introduce six ideation guidelines: defer judgment, encourage wild ideas, build on the ideas of others, one conversation at a time, be visual, and go for quantity. A full discussion of these concepts is beyond the scope of this text, but IDEO's *Design Thinking For Educators* provides a wealth of detail on this topic [77]. We have found these to be very helpful concepts that promote both a higher quantity and quality of team-generated ideas. We highly recommend starting with the "Yes, But/Yes, And" activity because it serves as shorthand that students can use to remind each other of ideation best practices. Later in the semester we would occasionally overhear students say something to the effect of "Stop being such a *yes but*; Change those to *yes ands*!"

Figure 4.1: Students generating ideas on white board with sticky notes.

After coaching students on best ideation practices, we practice ideation around the topic of drones. As a pre-class assignment we ask students to bring an additional three ideas for drone use that they would be willing to share with their team. This helps to seed the ideation process. We have tried a variety of ideation techniques during class, but one of our favorites is brainstorming with constraints.

In this activity we give each student a pad of sticky notes. The team stands at a wall and, as they have ideas, they write them down on the sticky note, say them out loud, and put them on the wall. The idea is to generate lots of ideas, group similar ideas, and capture it all on the wall.

After about 5 minutes teams usually start to run out of steam, we inject energy into the activity by adding additional constraints. For instance, our first constraint might be something silly like, "for the next 2 min all the ideas you generate have to involve drones and bananas." For the next set of constraints, we additionally challenge students that all ideas must be, for example, larger than a car. We repeat this exercise with several additional constraints and then end with another round of open idea generation. The point of these additional constraints is that it gets students to think in categorically different ways. While many of these ideas are totally unrealistic, some elements may work their way into more realistic designs or generate ideas that might prove feasible. The point here is to emphasize quantity over quality—giving them lots of options to pick from when selecting their final design.

4.2.4   CHOOSING A DESIGN

There are a number of ways to select from multiple design alternatives. Rather than use a standard engineering approach, such as a weighted benefit analysis, we bring in a tool from the business world. SWOT, which stands for Strengths, Weaknesses, Opportunities, and Threats, is a methodological approach to decision-making. It can be applied to small decisions, such as choosing from among a set of design alternatives, or big ideas, such as company mergers. Strengths and Weaknesses focus attention internally on the idea itself—what works and what doesn't—while Opportunities and Threats focus attention on the externalities of the situation—is the market ready for this kind of idea? Simplicity helps explain why this approach is so popular. Even if one has never heard of the SWOT analysis, after reading just these few sentences most readers will have a basic understanding how to use this tool. We spend about 15 min explaining the concept in class and give the students another 30 to do a preliminary SWOT on their idea. For more on SWOT see *An Essential Guide to SWOT Analysis* by Gomer and Hille [78].

We provide students with a mini lecture on SWOT analysis and give them a few examples of how the tool is used. We then ask them to pick one of the ideas they have just generated and, using SWOT, decide whether it might work. Performing a SWOT analysis helps students practice a sociotechnical perspective on idea selection—as it forces them to go beyond simply evaluating the technical merits of the idea.

Over the next few weeks—i.e., while we are engaged in the activities described in the previous chapter—students slowly work on refining their idea. This is accomplished through a series of short faculty check-ins that lead to an expectation that rough ideas are coalescing into something they can really work on in the exercises described on the following pages Infographics, Minimum Viable Product, and Pitching.

4.2.5   INFOGRAPHICS

We require students to create an infographic and short executive summary that explains their chosen concept. This assignment challenges students to distill their broad concept down to a few key ideas. It is also a great scaffold for their final pitches—many of the ideas students use on their infographics can be translated into presentations. This assignment is also closely aligned with the requirements for the Global Social Innovation Challenge. The infographics have three components: the problem landscape, the solution landscape, and their drone.

In the problem landscape section students explore the root causes of the problem they are trying to address. Who is affected by this issue? What's its size and scope? We ask students to examine the political, economic, social, technological, environmental, and legal hurdles that are preserving the status quo. The solution landscape exercise exposes students to current efforts to solve this problem. What has been tried or is being tried? What has worked and what has not? The point here is to help students to see that their work does not occur in a vacuum, but is surrounded by others—sometimes collaborators and sometimes competitors—working toward similar or competing goals. Lastly, we ask students to describe their drone-based solution. We challenge them to articulate why drones are the right technology to address the problem. To do this, students must explain how they improve on existing solutions. It is in this section that students present the results from their SWOT analysis. This approach is broadly matched on the Problem Landscape, Solution Landscape, and roposed Solution framework that anchors the Master of Arts in Social Innovation curriculum at the School of Peace Studies, where Austin teaches.

We also use the infographic to introduce a key idea: *rapid prototyping*. When engineers hear the phrase, they tend to think of 3D printing. While this is one application of the term, we require something far simpler in our class: students must simply sketch a wireframe of their concept. After introducing this assignment, we give the teams about 30 min and challenge them to come up with a plan for their infographic and draw a wireframe on the whiteboard (see Figure 4.2). We emphasize the importance of focusing on the big picture and deciding which key elements they want to present. After students complete their prototypes, we do a gallery walk: one team member stays with the infographic to explain it to visitors and keep track of feedback. The remaining members visit the other infographics. We ask students to critique each other's concepts at this very early stage—providing actionable suggestions for improving the infographic. This early feedback prevents students from spending time on less viable elements while also enhancing the quality of the final infographics.

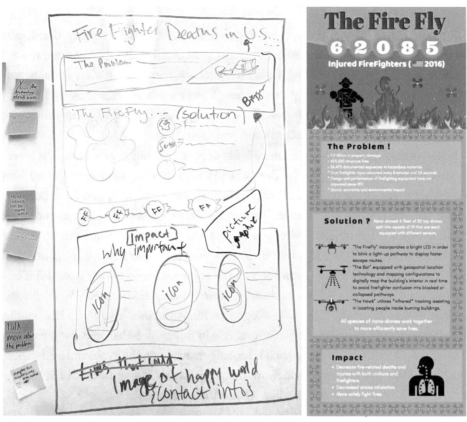

Figure 4.2: Infographic rapid prototype and final infographic.

4.2.6   MINIMUM VIABLE PRODUCTS

Unlike many engineering design courses, we *do not* want our students to build a full-scale engineering prototype of their drone. We do, however, want them to create something tangible to demonstrate that they are on the right track. Borrowing from the Lean Startup methodology, we have students develop a minimum viable product (MVP) [79]. Eric Ries, founder of the Lean Startup, defines the MVPs as "that version of a new product which allows a team to collect the maximum amount of validated learning about customers with the least effort" [80]. An MVP is explicitly *not* a prototype of the full system, but is instead something that allows designers to validate some of the assumptions inherent in their design. The classic example of a minimum viable product is a website landing page with a button that says "click here to order"—if no one clicks the button, there's no reason to build that product.

Applying this to hardware can require a bit more effort, but it is well worth the time. For example, one of our teams proposed using a drone to search for evidence of palm weevil infestations atop palm trees, a particular local problem in San Diego. At first the team thought that testing their system would require them to build a drone with a specialized camera and collision avoidance. After learning about MVPs, however, the team realized that their fundamental question was not whether a drone could fly over palm trees—which it obviously can—but how close the camera needs to be to the tree to detect a palm weevil infestation. Recognizing this as the critical assumption, the students realized there was a much simpler MVP that would let them test this assumption: attaching a digital camera to a long pole and taking photos of a palm tree from several distances (see Figure 4.3). Had the project continued, they could have shown these photos to palm weevil infestation experts and asked the question "Is this good enough?" Another team interested in building an drone based land mine detector strapped a metal detector to a bike to get a hands-on sense of how speed impacts detection. While MVPs are conceptually similar to low fidelity prototypes, the main distinction is that the entrepreneurship MVP is designed to test key assumptions both about the product and customers.

Figure 4.3: An example of a minimum viable product for a palm weevil infestation detection drone. This camera on a stick is dramatically simpler than building a fully integrated drone platform.

4.2.7   PITCH COMPETITION

The final deliverable for the class is a 6-min entrepreneurial pitch competition aimed at convincing the relevant stakeholders to adopt the team's idea for a drone for good. The target audience for these pitches varies depending on the topic: some teams gear their pitches toward a panel of investors, while others imagine they are presenting to the board of a foundation. We intentionally keep these presentations to a short six minutes: crafting a clear and cohesive pitch in a short time span is an incredibly challenging task. Once they've pitched their ideas, students field questions from the judging panel for another 6 min.

To increase a sense that their work matters—within and beyond the classroom—we invite working professionals to serve as our judges. This dramatically changes the experience for the students. We find that they take the activity much more seriously knowing that they will be presenting to a group of industry experts. We have had a range of folks serve as judges, including a retired U.S. Navy drone pilot, the founder and CEO of a local drone company, a faculty drone enthusiast, and the Assistant Director for the University's Center for Peace and Commerce.

Preparing for the pitches is a multi-week intensive experience that, coupled with the development of the MVP, consumes the last five weeks of the semester. During these weeks we spend the majority of the class period meeting one-on-one with the teams. In these conversations we review their progress on the MVP, help them craft their message and draft pitches, and work with them to develop interim deadlines to ensure they successfully complete the assignment.

To help students craft great pitches, we begin by emphasizing the ways in which a pitch is simply a story: the goal is to draw in one's audience, gain their trust, and convince them of one's position. In the PowerPoint era there is a constant tendency to focus on slides. Conversely, we emphasize that slides are the very last thing that students should be developing, and instead they should start by focusing on their message. The Message Map is one tool we have used to scaffold these messages [81]. In this activity we ask students to create a simple headline message about their drone for good, support it with three possible benefits, and back those up with key stories, statistics, and examples. We then have students repeat the rapid prototyping example and sketch out a visual storyline that supports their message. Only then do they begin developing the presentation in slide-deck form.

With two weeks remaining in the semester we have students do a full rehearsal of their pitch. Rather than present to the whole class, we reserve a separate classroom and have the students present to the two instructors and one other team. The observing team is instructed to take detailed notes and then provide feedback to their classmates. After the observing team delivers their feedback, they return to the classroom and we meet one-on-one with the team to debrief and give them notes. We have found this model to be extremely successful as it helps build students competency not only in pitching, but also in providing feedback to their peers about how to improve pitches. Often, we find that students make a mistake in their pitch, get some feedback from their peers, recognize it in the next presentation that they watch, and then fix it before giving their final pitch. This process certainly makes our job easier.

We hold the final pitches in a formal theater space in the Peace School to signal to everyone that we take the pitches seriously. Knowing the final pitches will be in a venue other than the classroom helps focus the mind, and the fact that the final event is in the Peace School—which half our students call home—sends a clear signal as well.

Attention to these details further differentiate this experience for the students from their normal classes and increase the sense that their work—whether in the classroom or beyond—has real-world applications and implications, i.e., is authentic. We invite our judges to ask probing questions that challenge students to consider their concept in greater depth and from a fresh perspective. We find the best teams have taken our advice and prepared backup slides that clearly address the judge's key points. After each of the teams have presented, we join the judges in a separate room and facilitate a conversation to select the pitch competition winner. We return to the theater, announce the winner, and then have a bit of pizza as students can unwind, debrief and, if they so choose, talk to the judges and learn more about their professional work.

4.3   CROSS-CUTTING THEMES

So far, we have described a course that emphasizes sociotechnical thinking in the context of drones. While these two objectives are most important, when developing this course, we realized that there was an opportunity to introduce a third element—what we consider to be a cross-cutting theme. In this section we briefly conceptualize these cross-cutting themes as modules that can be swapped out in subsequent versions of the class.

4.3.1   ENTREPRENEURSHIP

Entrepreneurial thinking is often mistaken as exclusively relevant to startups. Following the KEEN Foundation, we take a broader approach and define entrepreneurial thinking as helping students to see the bigger picture—specifically through what KEEN calls the 3C's: curiosity, creativity, and creating value [82]. *Curiosity* is about helping students develop a critical consciousness—fostering the desire to question authority in all forms and explore contrarian perspectives. *Creativity* is about thinking "outside the box"—bringing fresh eyes to intractable problems. *Creating value* emphasizes the importance of developing solutions that have both a financially sustainable business model and have a positive impact on the world.

Entrepreneurial thinking has been a recent theme in both our School of Engineering and School of Peace. It was not much of a stretch to integrate this course with the University's existing suite of social entrepreneurship efforts. In particular, the Social Innovation Challenge, described in the previous section, provided an excellent scaffold on which to build our course. Specifically, we incorporated entrepreneurship into the class through a number of key resources. First, we situated our efforts within the Lean Startup Methodology, with a particular focus on the Minimum Viable Product. Second, we experimented with the different resources provided by the KEEN Foundation such as their online module on pitching. While students disliked the format of the online modules, they encouraged us to integrate the content from those modules directly into the class—a suggestion we heeded and that was well received in future course offerings. Third, we emphasized the importance of creating a compelling pitch, as opposed to a piece of hardware, as the ultimate deliverable.

4.3.2   EMPATHY

In our first version of the class we anticipated that the notion of *Empathy* would be a valuable way to conceptualize the reason our multidisciplinary approach was both needed and useful. Our approach is rooted in Oxley's definition of empathy as "feeling a congruent emotion with another person, in virtue of perceiving her emotion with some mental process such as imitation, simulation, projection, or imagination [83]." This definition holds our attention because of its focus on both the affective (an emotional register between two people) and cognitive (a way of understanding how others see things).

Engineers are increasingly attentive to the importance of empathy, especially in the design context. Leading institutions like design firm IDEO and Stanford's d.school have done an excellent job highlighting the importance of a user-centered approach to thinking about the ideation, prototyping, and manufacturing of products and processes. We applaud that approach, and trust it will close important loops in the design and manufacturing process [77].

Our focus, however, is on the importance of empathy within pedagogy and practice writ large. A broad review of the literature, and interviews with engineers in industry and educators in the academy, has found that while empathy is considered to be important, it is rarely addressed explicitly in either industry or educational contexts [84]. Some of our colleagues are optimistic about the possibility for change, suggesting empathy is "a teachable and learnable skill, a practice orientation, and a professional way of being" in engineering [85]. We couldn't agree more.

We designed the first iteration of our course on the assumption that its interdisciplinary nature would lead to increased empathy. In that version we repeatedly signaled to students during class the importance of seeing things from another's perspective. We created time and space to talk about disciplinary differences and help the students critically reflect on the way in which their studies have impacted their world view. In one particularly effective exercise, originally developed by Barbara Karanian at Stanford, we asked students to think of a particularly emotional event (any kind of emotion was fine) and draw a picture of that time. We then paired students and challenged them to guess the emotion drawn by their colleague [86]. Using this exercise early in the semester helped spark conversations that brought students closer together and bridged the disciplinary divide. It also laid the groundwork for future conversations, in which we used the language of *empathy* to discuss one of the benefits of a sociotechnical and multidisciplinary approach.

The tricky thing about empathy, though, is that it is incredibly hard to measure. Nevertheless, we set out to collect in-depth data throughout one semester. Qualitative data took the form of focus group discussions, faculty reflections, ethnographic observations, and student work. Quantitative data took the form of pre/post student surveys and student work. We crunched all the data and wrote a few articles [13, 15, 16, 87], but at the end of the day our findings can be nicely summarized by one student comment:

> *"You know, talking with a lot of people in the class seeing how, when you talk to an engineer you get one perspective, and because of the way we've been trained that's usually the same. But when we talk to a Peace and Justice student, I'm like, 'Oh, I never thought about all these other possibilities'."*

Whether such student's realizations translate into empathy is something we, in the final analysis, found difficult to verify empirically. Moreover, it seems likely that, if it exists, empathy is a byproduct of the interdisciplinary process, rather than an outcome from specific programming. For this reason, we determined that a shift from a focus on empathy to a focus on entrepreneurship would likely generate empathy just as efficiently.

4.4   SOCIAL JUSTICE

Another option for a cross cutting theme is social justice. The decisions made by engineers and peace builders, in particular, shape whether social and economic relations are fair and just. We currently spend one week on this topic, but it could certainly be expanded to be a consistent theme over the course of the semester. Social justice is both incredibly important and extremely difficult to teach, especially in an interdisciplinary space. We scratch the surface and challenge students to simply define what social justice means—and even that is a nebulous conversation. As Riley explains in her book *Engineering and Social Justice*:

> *It is difficult to define the term social justice. It is not that the term is poorly understood; on the contrary, each of us knows what we mean by it. The problem is that the term resists a concise and permanent definition. Its mutability and multiplicity are, in fact, key characteristics of social justice* [88, p. 1].

Fortunately, there are a wealth of resources about teaching social justice writ large [89–91], and for engineering specifically [2, 88, 92]. We argue that sociotechnical thinking is a prerequisite for social justice work, both within engineering departments and beyond. We hope our efforts in this class lay the foundation students can build upon as they engage with more complex justice issues. When designing this class, we hoped the peace students would naturally raise social justice critiques as their interdisciplinary teams worked through the engineering design process. While this happened on some teams, we were somewhat surprised to see it not come up more often. As a result, we ended up playing that role in our one-on-one meetings. While a "Drone for Good" does not necessarily have to promote social justice, when teams proposed ideas that contrast with social justice commitments—by suggesting drones that monitor "high crime" areas without recognizing the systemic issues in policing communities of color—we stepped in to ask clarifying questions about the broader context and specific implications. We had thought Peace Studies students would have been overly sensitive to these issues. They were not. Why is this? We suspect our efforts to promote a *Yes, And* approach rooted in empathic and interdisciplinary engagement may have led some of our more critical students to suspend judgment or hold their tongue. Perhaps we should later reintroduce *Yes, But*, which an observant reviewer noted has a clear role in this process as well, especially when the class shifts out of brainstorming and into more critical explorations of technology's impact.

A laser focus on social justice in this course would give it a very different flavor. Sociotechnical thinking has proven to be helpful in aligning students from disparate backgrounds to a common cause. Our discussions of social justice, on the other hand, often surface substantial divisions between the engineering and peace students. Sociotechnical thinking provides a good bridge for conversations across disciplines, but it does not challenge student's fundamental assumptions about what a fair and just society looks like, and how to get there. As an aside, our colleagues have had good luck teaching engineering and social justice elective course exclusively tailored to engineers. In some ways these homogenous classes can be more effective as the social justice critique of engineering is coming from within, rather than a charge levied at engineering by the hippy professor from the peace school.

4.5   CONCLUSION

Thus far in this volume we have focused on how to use disciplinary differences within the classroom to promote student learning. In this chapter we set out to highlight what we feel to be one of the most unique benefits to this class, which is the opportunity to harness the course's unique format in order to pursue some particular objective. We have approached this project with the sense that having students working across differences generates a series of new experiences, but that there is enough capacity within this process to dedicate some time to building specific skill sets and sensitivities, whether it be empathy, entrepreneurship, or another topic of the educator's choosing. In other words, if the sociotechnical approach is a means to an end, we hope to have provided a framework that allows the instructor to choose their own ends.

For example, an interesting and engaging class could be organized around environmental sustainability (rather than drones for good), with the key projects drawing on the construction of air quality sensors (rather than building drones and prototyped uses), that are then deployed within a community exposed to pollutants (rather than each team's chosen project), with the entire class focused on the cross-cutting theme of citizen science and public policy (rather than empathy or entrepreneurship).

While the range of possible topics is broad, our experience suggests a number of factors will increase the likelihood of success. To be beneficial, cross-cutting themes should be:

- sufficiently broad to justify semester-long focus;

- sufficiently resourced to ensure appropriate course material is available;

- sufficiently balanced between the two or more groups represented in one's heterogeneous class;

- focused on an expertise held by the faculty teaching the class; and

- sufficiently linked to broader campus efforts and commitments.

No two educational institutions are the same, so there is clearly a wide range of contexts, resources, commitments, and experiences on every campus. These benchmarks reflect the lessons we've learned over the past few years, and we share them in the hope that they prove useful in subsequent iterations, regardless of the topic, project, or cross-cutting theme.

Astute readers will notice we've made it halfway through the book without saying much about drones. The next two chapters focus more directly on the technology itself and have been written in such a way that they can be assigned as reading within the class itself. In these chapters we focus on the complex interplay of the social and technical, providing first a tool for discussing the sociotechnical (Chapter 5) and then a set of case studies for trying out these conceptual tools (Chapter 6). Both chapters have been written in such a way that they can be excised from this book and assigned directly to students. As a result, these chapters do not refer to "this class" or to "students" in the third person. It has been written to be read by students. We hope this does not cause undue confusion.

CHAPTER 5

The Ethics of Drones

Microdrones by Randall Munroe (More at xkcd.com)

5.1 INTRODUCTION

The social and the technical are interwoven in very important ways. We believe this interweaving has been under-appreciated by our home disciplines of engineering and the social sciences. In this chapter and the next we're going to be playing with some of the implications of this simple set of observations. What if engineers (like Gordon) took the social, political, and economic worlds seriously when designing things? What if public policy people (like Austin) took more seriously the built world that shapes, constrains, and facilitates social action? Put another way, should engineers be thinking about ethics, and should social science folks be thinking about technology. You won't be surprised to know that we think the answers here are *yes* and *yes*.

This chapter takes a deep dive into the social and technical facts and implications of drone technology. We're focused here on consumer-level drones used for the greater good. What exactly are drones? How do we approach the ethical puzzles they create? We present a number of critical tools for engaging debates over how to balance this new technological capacity with social benefit. Our goal is to open debate that goes beyond drones, and in this way provides a viable framework for thinking about and working with emerging technologies. The role drones play in the sociotechnical ecosystem is not unique—any emerging technology brings with it extensive policy debates and lingering concerns within the general public. The Internet, for example, has extensive social, economic, and political implications that we are only now beginning to fully comprehend. As you read this chapter, we encourage you to take notes whenever you see something you don't agree with. That's the stuff good class discussions are made of!

5.2   WHAT ABOUT DRONES?

Drones, specifically the quadcopter variety we focus on in this chapter, have primarily been made possible by technological advances in mobile telephony. Customer demand in the smart phone market led to the rapid advances of accelerometers, gyroscopes, microprocessors, batteries, and wireless infrastructure. The explosive growth of the smart phone market has also meant that billions of dollars have poured into research and development—leading these technologies to get better, faster, and cheaper. Take MEMS accelerometers for example (MEMS stands for microelectromechanical system). This technology senses the orientation of one's smart phone and rotates the screen between portrait and landscape mode. MEMS were first developed for intercontinental ballistic missiles and cost thousands of dollars—at the time it was thought that might be the only viable application for this technology. Now they cost pennies. These technologies are one of the things that make smart phones smart. As it happens, they are also the core element that makes stable drone flight possible.

So, what about drones? Are they sufficiently unique to merit sustained attention? We think the answer is a resounding *yes*, and we think this for three reasons: first, they allow for new activities to take place in the air; second, these activities are subject to greater levels of technical command and control; and third the process of acquiring and using these devices has been democratized [12]. Let us take each of these transformations in turn.

First, drones allow new activities to take place in the air. A wide range of innovative efforts by hobbyists, specialists, and commercial interests have begun probing the many ways drones can be used to enhance the existing efforts of business, government, and civil society groups. Drones extend these groups' abilities beyond what had previously been possible with existing technology. This use is sometimes *emergent*, by which we mean it cannot reasonably have been done previously without the resources of the nation-state or similarly large and well-resourced institutions. Occasionally this use is also *disruptive*, by which we mean that drones make it possible to do things that would simply not be possible without this technology [10].

Second, more sophisticated control systems have made non-military and non-commercial flight safer and easier than any previous small-scale flight platform. Drones from Chinese manufacturer DJI, for example, can fly themselves. A host of sensors also makes it possible for these drones to easily follow their owner while avoiding obstacles along the way. These sophisticated control systems have been combined with lighter weight and longer-lasting power systems that make extended flight times possible. On the horizon we see a new generation of networked capability that ensures the stable and reliable interface of drone systems with other systems, including radar, air traffic control systems, real-time swarming, real-time collision avoidance, non-radio frequency real time control systems, and so forth.

The third sociotechnical implication we identify is the democratization of this technology. More stable and reliable systems available at a lower price point has meant that a much wider range of groups and individuals have begun experimenting with drone systems. While surveillance from the air was previously limited to government agencies that could afford aircraft like the U2, now activists can acquire a drone for a few hundred dollars, fly over physical barriers enacted by their government, and expose corruption (or do harm) in previously unimaginable ways.

5.3   SOCIAL IMPLICATIONS

A second social implication involves the safety of people and objects on the ground. Significant concern about airworthiness plagued early platforms, as glitchy GPS systems led to the occasional, but widely reported, fly-away of drones, especially of the industry-leading DJI platform. (Fly-aways refer to drones that have lost connection with their ground-based operators and are quite literally flying without any human control.) Subsequent innovation may reduce these concerns, as advances are made in drone detection, airspace denial, kinetic impact reduction (from external frames and parachutes, for example), and collision-avoidance systems (visual sensors, mesh communication). It remains to be seen whether these innovations will assuage concerns over safety. These are the challenges related to the platform regardless of payloads. Payloads themselves, however, represent a distinct safety concern. Fear that drones might be turned into flying improvised explosive devices (IEDs) are legitimate, as small platform drones have already been weaponized and projected over Israel's wall surrounding Gaza, for example. Drones themselves can be dangerous if they crash, and even more dangerous if they carry weapons.

A third social implication involves privacy. Drones upend a series of assumptions about whether particular places are public or private. Residents in a penthouse, for example, may have different assumptions around privacy (and their subsequent need for curtains) than do their downstairs neighbors with homes on the first few floors. It may seem like a silly example, but penthouse suites, the sunbathing rooftop, and the White House are each locations where their value is implicitly linked to their inaccessibility. Drones equipped with cameras, sensors, or proactive payloads fundamentally upend the link between location and privacy that had been established in the era of the land-based camera systems. Sure, it was previously possible to fly a helicopter over a hard-to-reach place, but not everyone had the means or desire to do so. Anyway, if you got into trouble a helicopter is very hard to hide, both legally and physically. Affordable, anonymous, and easy-to-fly drones have democratized this accessibility, and raised new concerns about privacy [93]. The point is clear, it is hard to stop people from using drones to spy.

We use the term *sociotechnical* to highlight the extent to which these technological innovations and the sociological implications intersect with one another. For example, new control system capabilities make it possible for multi-rotor drones to hover in place. This loitering capacity is the feature set that causes the most concern on the ground, namely, the fear that a nearby hovering drone is watching everything, or, more concerning of late, is only watching *you*. New sensor systems, including facial and gait recognition technology, make it possible for drones to follow a particular person as they try to walk away from it. The possibility that the drone is not watching anyone in particular, does not actually have a camera onboard, and is also not carrying any other sensor systems, is not the first or second thing that goes through the mind of private citizens in public places.

Drones' impact on perceptions of privacy is an important sociotechnical issue. In addition to their ability to see things in new places, they also create incentives to do new things on the ground, which can now be seen from the air. For example, Austin was contacted by activists interested in protesting a decision to bulldoze a community in order to put in a parking lot. The activists hauled stones into a nearby field, arranged them in the shape of a Parking sign that could only be viewed from the air, and had Austin's team fly a drone over the guerrilla art installation. The art was impossible to see without the drone—indeed, the art was built for the drone and because of the drone (Figure 5.1).

Drones do new things in the air, but also create new spaces for things to happen on the ground. All of these are technical uses with social impact, or social uses made possible by technological innovation. The two are inextricably linked, as the field of science and technology studies has long observed [21, 23].

Figure 5.1: Political symbols on land meant to be seen from the air (Austin Choi-Fitzpatrick and Tautvydas Juškauskas).

5.4   UNSETTLED TIMES

While an initial flurry of innovation has given way to a stable set of uses within the military space—think about the thousands of hours of flight time that the average Predator sees per-year while in support of military objectives—commercial and small platform drones are still a ways away from regular and routine deployment in broad commercial space. One reason for this is that a reliable and steady use-case has not been found. The reasons for this are social and technical.

If the commercial and small-platform space is undergoing rapid technological innovation, and if the socio-political issues have not been solved—then clearly the dust has not settled in three key areas: (1) *Use*: what should we be doing with these things? (2) *Regulation*: what's the right-fit public policy that balances innovation with other factors? and (3) *Public opinion*: should use and regulation push or follow public sentiment about new technology?.

Use: The rate of adoption is dramatic and the range of users and use is broad and heterogeneous (more about this in the next chapter). In our search for cases of purposeful and non-violent drone use, we were surprised by how rarely we found people using drones to break the law, and how frequently we found people and institutions using drones in scatter-shot and one-off experimentations focused on exploring the technology's applicability to a particular field. This range of experimentation continues into the present, and has yet to generate a robust set of uses in the social sector. Early experimentation in agriculture, earth observation, and arts and entertainment appear to have paid off as business have formed to scan and spray fields, monitor and inspect heavy industry, and provide affordable aerial imagery for the entertainment industry [9]. This leads us to anticipate that stable and settled use will continue to focus in industries where there is less of a need to overfly heavily populated areas, and a greater likelihood that established and resourced industries turn to drones as an affordable alternative to helicopters, airplanes, or satellites.

Regulation: A regulatory consensus appears to be converging on a set of requirements, in particular that: drones fly below 400 ft, within the operator's line of site, with the permission of local authorities, and with the proper registration of the device as well as the licensing of the operator. In theory this is the approach advanced by the Federal Aviation Authority, though regulations have taken longer to draft, much longer to implement, and have proven to be nearly impossible to enforce. At the state and municipal level policymakers have drafted a range of supplementary policies intended to restrict the use of drones by the police or by civilians. Internationally, a similarly wide range of policies have been adopted as legislative bodies debate whether to prohibit personal drone use, restrict them in some way, or allow unrestricted use. Across all regulatory zones concerns focus on security and privacy. Clearly, there is much work to be done in the regulatory space.

Public opinion: Public opinion appears mixed on the use of drones. A 2017 study from the Pew Research Center found Americans said they would be curious (58%) or interested (45%) if a drone flew near where they live [94]. At the same time, more than a quarter (26%) said they would feel nervous if a drone flew near them, with 12% suggesting this would make them feel angry and about one tenth (11%) saying they would feel scared. This may explain why so many respondents suggested drones shouldn't be allowed to fly near homes (54%), near accidents or crime scenes (53%), or at concerts, rallies, or other public events (45%). The most interesting result of this study, however, is the fact that younger people report feeling far less anxious about the technology than do older Americans. We believe acceptance will increase as novelty wears off, but this is just a hunch.

Our argument is that now is *exactly* the right time to take a sociotechnical approach to debates over drones. Unsettled times are perfect for asking *What can we do? What should we do? Should we do it just because we can? Should we not do it just because it's not currently acceptable? How should we navigate between feasibility and acceptability?* Engineers and change-oriented folks in the social sciences confront these puzzles all the time. The next section is all about navigating those tradeoffs.

5.5   ETHICS OF ACTION

We believe the puzzles raised by technologically unsettled times, and thinking critically about the cases we raise in the next chapter, require an *ethics of action*—a set of standards to use when evaluating drones. While the FAA has set out a series of policy requirements, non-commercial actors are largely left to their own devices when it comes to responsible use. What norms should guide this use?

It is important to note that a broader code of conduct has yet to be agreed upon. Critical, we argue, are a series of ethical approaches that attempt a balance between innovation and other concerns, in particular safety and privacy. In the following six points we draw extensively from Austin's earlier work [8, 12]. Readers are invited to reflect on whether this list is appropriate, whether the threshold levels within each item are properly set, and whether a single series of prescriptions can be applied more broadly.

1. *Subsidiary*: Should drones only be used in those situations where other actions or technologies already yield the desired result? Can new technology be original without being useful? If so, how might we know the difference?

2. *Physical and Material Security*: Appropriate measures (training, flight-planning, equipment checks) must be taken to ensure the security of people and things in the area where a UAV is used. As drone use increases, who will coordinate these efforts? How will anti-establishment actors (e.g., protestors) fit into this space?

3. *Do No Harm*: This concept, pioneered by Patrick Meier and colleagues, emphasizes the importance of the public good: benefits must outweigh costs and risks. Yet the nature of the public good is a matter of great debate; does documenting an embarrassingly small turnout on a key social issue harm the movement's cause?

4. *Newsworthiness*: This concept is borrowed from journalism's focus on the public good and emphasizes the importance of a free press (in both corporate media and citizen journal models) in holding the powerful to account. Must advocacy footage only be made "for the greater good" or is aerial data collection important in its own right? Is ubiquitous aerial image capture a simple step up from Google Earth in terms of frequency of coverage or is it a scale shift that represents a fundamental threat to privacy?

5. *Privacy*: While debates about privacy and technology are ongoing, and users of digital media appear less worried about the issue than advocates (at least in the U.S.) what is the proper balance between the privacy of private citizens on one hand and newsworthiness and the public good on the other? Privacy is treated differently across national contexts, and no blanket legislation is possible, meaning the increased use of drones is likely to lead to very different policy approaches.

6. *Data Protection*: Data protection is critical. Social movements using camera-equipped drones to monitor police action at a political protest, for example, must take great

care to ensure that the privacy of protestors is protected and that the digital data is kept secure. The inverse is true, as police surveillance data should also be subject to Freedom of Information Act (FOIA) requests and public accountability.

It will be immediately obvious to the reader that some of these criteria are in tension with one another. How do we strike a balance? Perhaps a case study will highlight the utility of these principles as well as the tensions inherent in their application.

5.6   APPLYING THE ETHICS OF ACTION TO A CASE STUDY: HUNGARIANS SPEAK OUT

In 2014 Austin and a team of graduate students set out to use small quadcopters to document the size of protest events in Budapest, Hungary, where he lived at the time. Together the team documented events large and small, and their activities culminated in the documentation of two of the largest protest events in the country's post-Communist history.

Viktor Orban, the authoritarian Prime Minister of Hungary, had passed a law that would tax Internet usage at the rate of one dollar per gigabyte. Orban's regime had drawn significant attention for its hostile approaches to the European Union, labor laws, women's rights, minority rights, immigration, and civil society in general. Through a combination of widespread public apathy, clever marketing, and the consolidation of the free press into the hands of a few regime cronies, consistent efforts to model Hungary more closely on Russian-style authoritarianism had been met with only small-scale protests. The Internet tax, however, infuriated both the left and the right, as well as the old and young, alongside both anarchists and businesspeople.

Turnout for the first anti-tax protest was huge, and tens of thousands of people took to the street. The Orban regime, however, used their media might to suggest the events were the same sad events attended by social outliers and misfits. Unbeknownst to the authorities, Austin's team had filmed the entire event, collaborated with an independent media outlet, and released the footage online. It quickly went viral, racking up tens of thousands of views within a few hours. A second protest was mobilized, and it was even larger. Austin's team made a second video with our colleagues, and it was similarly popular [95, 96].

The government dropped the policy in short order and news of the retreat made the front page of the *International New York Times*, alongside a photograph of the protestors pointing their illuminated phone screens up to their drone. A short time later the team was invited to a government-hosted conference on the use of drones. A draft policy was circulated, no doubt in response to our efforts. Among the stipulations: *drones could not be flown over crowds*.

Austin's team went on to develop a method for estimating the size of a crowd. They used grids and density estimates alongside basic ratios for determining the relationship between a digital image made from the air and the associated space on the ground. The technical approach won an award and Austin and his team at the Good Drone Lab produced a number of additional articles on the topic [9, 14].

5.6.1   DISCUSSION QUESTIONS

1. Does this usage match our ethical guidelines (Subsidiary, Physical and Material Security, Do No Harm, Newsworthiness, Privacy, Data Protection)?

2. What tradeoffs exist in this case, and are they worth it?

3. What challenges exist?

4. Would these challenges be ameliorated by legislative approaches, technical adaptation, social change, or some other approach?

5.7   APPLYING THE ETHICS OF ACTION TO A REAL CASE

As this case study shows, in 2014 there was little legislative oversight for this kind of use, whether in Europe or the U.S. As a result, Austin developed his own ad hoc guidelines. In this section we model one approach to discussing this first question: *does this usage match on our ethical guidelines* [8, 10].

Subsidiarity: Is it possible to estimate the size of medium to large crowds using existing approaches? At present, there is no auditable method for estimating the size of a crowd in an unbounded space. By sealing off a space and adding a turnstile, one can easily measure ingress and egress—but this violates the *unbounded* space condition that applies in most public events. Use estimators to count off through the crowd in a rigorous way and anyone can generate an estimate—but this violates the *auditable* and *affordable* conditions that makes this approach apolitical. We determine that we have met the subsidiarity threshold.

Physical and Material Security: We did our best to launch, fly, and land our UAV beyond the edge of the crowds depicted in our work on crowd estimation. New parachute technologies have emerged in the time that has transpired between our data gathering and this publication. These would add a further level of safety to our flight. Furthermore, Austin's team has tested camera-equipped balloons which remain tethered to the ground, thereby eliminating a host of safety concerns. We determine that we have nearly met the security threshold in our past efforts, but future efforts will certainly meet them fully.

Do No Harm: The gathering and publicizing of data about public events is inherently in the public interest and the provision of this data is for the public good. Our activities could have caused harm had our camera captured individually identifiable faces. Critics have pointed out that the current analysis overlooks another significant area of potential harm: our footage could highlight the extent to which important events suffer from low turnout rates, thereby amplifying criticisms from opponents. We thus leave open the question of whether Austin succeeded in doing no harm.

Newsworthiness: Gathering and publicizing data about events that social actors desire to make public are inherently worthy of public attention. As a result, their documentation of a protest event is decidedly newsworthy. Whether their documentation of private citizens at a public concert is newsworthy is less clear-cut, although we feel that such events are regularly covered by newspapers in the arts and entertainment section. We leave open the question of whether Austin met the newsworthiness threshold in one of our two cases.

Data Protection: Data captured during public events should be secured. How it is secured, and at what level of protection, is a matter of ongoing debate. All of the raw footage for this project is stored in a Dropbox account, which syncs over password-protected WiFi connections to the hard-drive of a password-protected MacBook Air. Is this a secure arrangement? This approach is sufficient for apolitical data, but would be easily hacked by a sovereign, or state-sponsored agents intent on disrupting protest activity. Such data protection is sufficient at one level and insufficient at another.

Privacy: By only engaging the camera function on the drone at a high altitude, Austin elided the complicated issue of privacy. Activating the camera at a lower altitude, however, was technically feasible and would have certainly captured discernable faces. Here we find a puzzle: should activists document public events in such a way that capture individually identifiable features? To date, citizen journalists have argued in the affirmative, and a wave of scholarship on new digital technologies like smart phones has suggested that these new tools level the playing field when it comes to capturing and telling stories [97]. Our sense, however, is that individuals who express enthusiasm for smartphones are often more sanguine when it comes to drones equipped with cameras or other sensors. Why might this be? A sustained conversation about the deployment of drones by protestors, police, and the media is long overdue, and will raise far more questions than we might answer here. Returning to this study, we have respected the privacy of individual actors by capturing and presenting data that obscures individual identities.

We will let the reader determine whether or not Austin and the team met these thresholds. More broadly, we hope these guidelines are subjected to real debate, as they represent an initial effort to establish broadly applicable ethical norms. Our thinking is that these could guide individuals and institutions in establishing specific guidelines around questions like (1) who gets to fly these devices?, (2) where?, (3) with what training?, and (4) under what conditions?

5.8   CONCLUSION

In summary, a sociotechnical approach helps us to better identify the interplay between technical innovation and social implications. We hope to have shed some light on the iterative nature of this relationship—it's not as if new technologies are invented and then dumped onto societies, nor is it true that societies move along without being affected in any way by new technologies. Anyone paying attention to anything knows this, but it's unusual to have these debates in the classroom. Both of us had to wait until we left school and entered our first jobs—Gordon for a defense contractor and Austin for a human rights group—before we learned this for ourselves. That's why we think the best way to bridge these gaps is through case studies that highlight both the technical and social dimensions of any particular issue. The next chapter focuses exclusively on a number of cases that we hope will start meaningful conversations about the sociotechnical.

CHAPTER 6

Drone Use Case Studies

6.1 INTRODUCTION

This chapter presents select case studies that explore many of the common issues associated with drones. We believe case studies are a perfect way to identify and discuss the complex relationships between technology and society. What rules should govern new technology? Should these rules be formal laws or informal social norms? Who should decide? We wrote this chapter because we believe these are important debates, and that practicing them—in the classroom or over drinks—will make society better. If you're a student reading this chapter, you should know that our primary goal as authors is to spark conversations about the tensions between social and technical systems. We chose drones because we think they are fun, but also because they raise challenging debates over (1) what's technically possible and (2) what's socially desirable. We are writing in the U.S. where the answer to both questions is *who knows*?! Now is a perfect time to be having big debates over big ideas, especially in relation to technology and society.

Each case will present a particular use of a drone and pose discussion questions that surface important tensions between technical, political, social, ethical, economic, and justice dimensions. This approach introduces specific puzzles and highlights key questions for discussion. We hope these cases will draw out meaningful debate between different voices and perspectives. These case studies were selected from within a comprehensive dataset of more than 15,000 reports of drone use from 2009–2015. Austin and his team at The Good Drone Lab scraped the web, blogs, and reports for purposeful and non-violent use of drones and identified more than a thousand discrete uses. Their analysis of those data led them to the realization that drones were being used by a surprisingly wide range of actors (Figures 6.1 and 6.2; note: all instance were only counted once, even if there were multiple reports of that use).

Furthermore, they found these devices were being used for a surprisingly wide range of purposes.

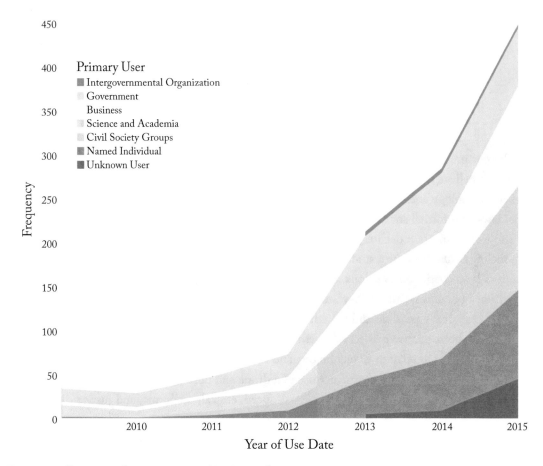

Figure 6.1: Drone use by primary user (2009–2015).

The full report and dataset are publicly available,, and their findings emphasize the extent to which the technological field is unsettled [14]. All evidence points to the possibility that UAVs will constitute a hothouse of research and development as well as an unsettled space for debate for some time to come. In other words, there is an open question regarding which of these uses will survive into the next five years. That's what makes our case studies ripe for debate.

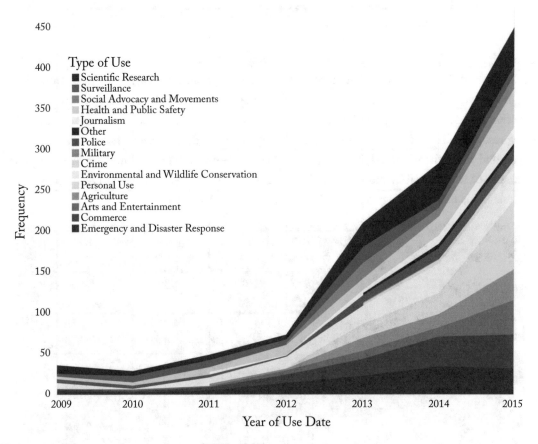

Figure 6.2: Drone use by primary use (2009–2015).

6.2   CASE 1—WHALES: RESEARCH APPLICATIONS FOR DRONES

Looking at the effects of climate change and growing food shortages in the ocean, Jan-Olaf Meynecke worried that the health of humpback whales might be suffering in ways unseen from shore. The humpback whale population has increased dramatically from their formerly threatened status in recent decades, however, growth in their population has forced the species to compete for dwindling resources. To better understand the effects of these trends, Meynecke hovers a drone over a whale, and when the whale exhales through its blowhole—twice every 3–4 min—a drone captures a sample of mucus within the exhaled air. Meynecke then brings back these DNA samples for analysis in his marine biology lab at Griffith University in Australia.

Figure 6.3: Researchers are increasingly using drones to study whales. iStock image.

Before drones, this type of research was incredibly difficult because the act of sampling DNA required the use of boats and crew, which disturb the whales' behavior. Additionally, boats and crews are expensive and intrusive. Alternative aerial sampling methods, such as the use of large remote helicopters, is not only expensive but also more dangerous.

"The fact that drones have become more affordable and easier to control, with more air time, provides a completely new dimension for research," Meynecke told our research assistant, Elizabeth Cychosz. "We might soon be able to even collect skin samples and attach sensors to them." Whales are only one of many animals that drones are now able to approach. Drones are increasingly used to document wildlife, raising fresh questions about machine-animal interaction, including whether drones induce stress in the observed species [98–101].

Scientists in fields as varied as archaeology and meteorology have adopted UAVs as a tool for collecting data from previously inaccessible locations. The incorporation of drones into these projects has enabled scientists to conduct studies that would have formerly been unsafe for humans to conduct themselves. Scientists run into many of the same legal hurdles as other drone users, with local legislation often limiting where they may conduct certain types of research.

Technical concerns may also limit what types of scientific research are possible. Although drone technology is becoming more efficient, powerful, and affordable, research designs are still limited by what drones are able to do at this point in time. For example, drones are still too intrusive for Meynecke to collect humpback whale skin samples. Our students have proposed using drones to tag sharks and to tranquilize elephants. These efforts are still in their early stages, but prompt us to wonder *how should we decide whether to continue research in this area?* Innovation in this space interacts with legislation, as laws and policies—e.g., no fly zones—could support or prevent ecological research.

6.2.1   DISCUSSION QUESTIONS

1. Does Meynecke's work match on our ethical guidelines (Subsidiary, Physical and Material Security, Do No Harm, Newsworthiness, Privacy, Data Protection)?

2. What tradeoffs exist in this case, and are they worth it?

3. What challenges exist for the researchers, the public, or the whales?

4. Can these challenges be ameliorated by legislative approaches, technical adaptation, social change, or some other approach?

6.3   CASE 2—GATWICK: ROGUE FLIGHTS OVER SECURE AREAS

While we were working on this book, multiple reports of drones flying near Gatwick Airport generated a wave of runway closures, flight cancellations, and public frustration in London [102–106]. With the first drone flight sighted on December 19, 2018, and repeated sightings for the next two days, authorities cancelled additional flights and began an investigation. The flights were reported by 115 people, 93 of whom were determined to be credible sources. In the course of the investigation, however, Sussex Police were unable to secure evidence that any drone had actually flown. The Royal Air Force (RAF) was called in, as UAV sightings continued to pour in.

Early reports suggested that the airport might soon reopen after the deployment of a "Drone Dome" sold by an Israeli weapons manufacturer. These reports proved faulty—it turns out the system had been ordered—likely after two drone-related incidents at Gatwick a year earlier—but had not yet been delivered. Once the RAF arrived, they set up a temporary system of their own and maintained it during their involvement in the incident.

Figure 6.4: Some airports have experienced substantial disruptions due to unauthorized drone flights. iStock image.

When the dust had settled, around 1,000 flights had diverted or canceled, impacting approximately 140,000 passengers. It is unclear whether the Drone Dome would have prevented the situation from occurring, as the RAF's replacement technology did not seem to stop fresh reports of drone sightings from pouring into the operation. At the time, the Gatwick's Chief Operating Officer stated: "We have had the police, we have had the military seeking to bring this drone down for the last 24 hours and to date that has not been successful." Two UAV hobbyists living near the airport were arrested, but neither was charged. As we write this case study many months later, it is still unclear who was flying these devices and if any such flight actually took place.

6.3.1   DISCUSSION QUESTIONS

1. What is the fundamental problem highlighted by this example?

2. What do you think the authorities should have done?

3. Where should we look for solutions to this problem (legislation, technology, social, process, elsewhere?)

4. Who should be held responsible for the disruption caused to the 140,000 passengers?

6.4   CASE 3—PUBLIC AIRSPACE: DRONES EVERYWHERE! MANAGING UBIQUITY

Drones hold the promise to radically disrupt our everyday lives. They are being adopted by firefighters to more quickly survey burning buildings, police to provide rapid responses in unknown situations, and Amazon to more quickly deliver one's packages. Where, exactly, will all these drones fly?

We suspect people feel entitled to a view of the sky, that this is an expectation—to look up and see the sky—shared by people regardless of political affiliation or socioeconomic status. We are writing from California, where significant resources have been invested to place utility lines underground and where high-rise apartments face coordinated resistance whenever they block the views of existing buildings. A committed appreciation for open skies is so widespread that it is baked into American colloquial tradition, as evidenced by the connection between freedom and phrases like "blue sky thinking" (to let the imagination roam unbounded) and to "reach for the sky" (to let one's ambitions stretch). The sky is a social and political metaphor for freedom.

Figure 6.5: Drones are increasingly making their presence known in the public airspace. iStock image.

It is in this space, in the skies that are humanity's metaphor for freedom, that drones will fly. But where *exactly* will they fly? One possibility is for them to fly in dense arteries over major terrestrial corridors like highways and byways. This would concentrate their deployment along areas where debates over where to place major infrastructure have already faced and overcome NIMBY (Not In My Back Yard) resistance. What's another artery, this one in a corridor 50 ft above traffic. Furthermore, the vast majority of the vehicles that ply these roads are hardened against the risk of a falling drone. Insurance for motorcycles and convertibles will likely go up, and there will be gawkers at first, but if self-driving cars and highway-based drone routes launch at the same time, their collision-avoidance systems can be integrated. If a drone drops a package on the road ahead of you, it may be that your car's computer has already received a message from the overhead drone that the payload bay door has failed, and an AmazonNow package is about to land fifty feet in front of you. The car's computer then automatically brakes faster than your senses can register the package, and traffic behind you (also in the self-driving grid) also comes to a halt.

Drones could also end up flying elsewhere—using alternate arteries, or perhaps in random patterns facilitated by collision-avoidance systems. Whatever the case, a network of sensors, combined with sophisticated control systems, are critical to ensuring any such system is deployed with minimal impact on the public and environment. A utopian view of this future evokes thoughts of the Jetsons; the dystopian approach summons *The Matrix* and Skynet.

6.4.1   DISCUSSION QUESTIONS

1. Should UAV-based infrastructure scale? If so, how and where?

2. What infrastructure configuration best matches on our ethical guidelines (Subsidiary, Physical and Material Security, Do No Harm, Newsworthiness, Privacy, Data Protection)?

3. Who should oversee this process: society, the market, the state, artificial intelligence, or some combination?

4. What tradeoffs exist in this case, and are they worth it?

5. What challenges exist?

6.5   CASE 4—WEAPON PLATFORMS: FROM GNAT TO PREDATOR

When people hear the word "drone," many think of the General Atomics' Predator platform, armed with guided missiles as it surveils and targets enemies of the state in an ongoing "War on Terror." Our colleagues in the human rights world have written many books arguing that the use of drones should be brought under the formally agreed upon rules of engagement that govern international conflict [107–109]. Others have gone further and called for a global ban on "killer robots," by which they mean weapon systems that identify, target, and kill without a human in the loop.

The man credited with inventing the predator is Abe Karem, and he recently sat down for an interview with Austin about his technology. We were interested to learn that Karem's original prototype was called the Gnat, and its first real job was flying for the CIA during NATO operations in Serbia. The international community was having a hard time proving that Serb forces were committing war crimes. Bosnian men were targeted and killed in an ethnic cleansing sweep focused on eliminating Muslim citizens, and their bodies were buried in mass graves when there was cloud cover, or at night, two times the Serb troops knew American satellites couldn't track their activities. Karem's drone prototype, which had long struggled to find a viable operational application, was dusted off and finally put to use. It complemented satellite imagery to prove war crimes were afoot—a remarkable achievement. With that intervention the cat was out of the bag and the Gnat began its evolutionary trajectory, eventually emerging as the *Predator*—itself the forerunner of the much larger Reaper.

Figure 6.6: General Atomic's Predator drones more closely resemble traditional aircraft than quadcopters. iStock image.

Fast forward to the present day and the Predator and its cousins are in heavy rotation as the United States continues to prosecute a global war that has pushed and broken international rules and norms about engagement [110]. The Predator pushes norms around weaponry in ways waterboarding pushes norms around interrogation. The one big difference is this: there is international consensus that waterboarding at Guantanamo is torture [111], but there is no international consensus about how to regulate "unmanned" weapons platforms.

6.5.1   DISCUSSION QUESTIONS

1. Should weaponized and remotely piloted craft be regulated differently than their Human-on-Board equivalents?

2. Who should oversee this process, society, the market, the state, or artificial intelligence?

3. What tradeoffs exist in this case, and are they worth it?

4. What are the hurdles in implementing your desired oversight process?

6.6   CASE 5—NEW LAWS: STOP FLYING THAT DRONE IN (MY TOWN)!

"If I wait on the feds, I'll be an old man by the time anything happens," points out Steve Vaus, the mayor of Poway, CA, a town that was among the first to pass a drone-related law at the local level. The law was passed in response to the presence of small drones in and around wildfires. The devices limited first responder and firefighter's ability to address the fire, since it made it unsafe for helicopters to safely approach the vicinity.

Figure 6.7: Local governments are increasingly passing their own drone regulations. iStock image.

Poway's legislation bans the civilian use of drones whenever the city's Director of Safety deems it necessary. In speaking with Austin's research assistant Jean Paul Digens about his decision to pursue the ordinance, Vaus addressed drone users as well: "The safety of our first responders takes precedence to your hobby." Poway's experience is not an anomaly, as civilian-controlled drones have frequently interrupted emergency response operations in recent years. Of course, the strength of drone regulations that rely on activating "no fly zones" in pre-ordained portions of land—a sensible, safety-oriented approach in the advent of a wildfire, for instance—will only be realized if surrounding communities cooperate.

New policies from the FAA run alongside local laws like Poway's, and due to a rapid increase in drone flights it is likely that sub-state legislation will continue to grow alongside increasingly comprehensive state and federal law. It is unclear how disparities between these two regulatory zones will be reconciled. Poway's legislation, for instance, allows a city employee to ban drones from certain areas for a wide range of reasons, including "civil unrest."

It is possible that municipalities follow Poway's lead by establishing regulations that serve their community's specific needs. If frameworks such as Poway's are abused by law enforcement agencies, civilians' ability to use drones and other emerging technologies in ways that can benefit their communities will be hampered. While loopholes are inevitable in new types of legislation, laws such as Poway's run the risk of undermining established civil liberties depending on how they are enforced.

6.6.1   DISCUSSION QUESTIONS

1. What are the major advantages and disadvantages of regulation at the local level?

2. Where should the authority lie to regulate drones (city, state, federal, other)? Why?

3. How might conflicting regulations impact the use of drones (for good or evil)?

4. What approaches might be successful in discouraging the public from inappropriate use of drones (such as flying near wild fires and disrupting fire prevention efforts)?

6.7   CASE 6—POACHING: PROTECTING WILDLIFE WITH DRONES

Conservation efforts are increasingly turning to drones to protect endangered species. A small operation in Kenya got started by testing whether chili pepper-equipped drones might scare elephants away from areas where they are prone to poaching [112]. Drones are also being used by Sea Shepherd to combat Japanese whaling missions, and illegal seal hunting [113].

Figure 6.8: Drones are increasingly being used to protect endangered species. iStock image.

Larger organizations are getting involved as well. The World Wildlife Fund was recently awarded a $5 million dollar grant from Google for innovative drone-based approaches to conservation issues, particularly poaching. Their drone, able to fly for an hour and cover a distance of 18 miles at an altitude 650 ft, can expand the battle against poaching in Africa and Asia by providing information regarding animal locations, danger zones, and ranger deployment. Incentives to create high-tech solutions to challenges such as poaching allows conservation groups to level the technological playing field, particularly as prices by weight for elephant ivory and rhino horn far exceed that of gold, driving poachers to become more technologically sophisticated [114]. Working together, these efforts have had a significant impact on poachers' ability to function, while also leading to an increase in arrests of potential poachers. In some cases, the areas where drones are being used have completely eliminated poaching attempts.

Although these technologies may stem illegal hunting, an initial round of legislative challenges threatened their use. Kenya, a poaching hotspot, briefly instituted a broad ban on drone use, effectively halting such anti-poaching projects. Such regulatory measures are far-reaching and may affect well-intentioned projects aimed at achieving what seems to be a positive outcome. With such a lucrative black market, however, the disruption of these criminal networks can have an outsized impact on corrupt politicians and policymakers, particularly in economically marginalized countries where economic opportunity often comes from elected positions.

Anti-poaching operations have a new element of surprise. Expanding technological capacities greatly enhances the scientific and ecological understanding of animals. As drone technology spreads, so will innovative ways to protect endangered animals. Public policies respond, sometimes quickly and sometimes slowly, and with complicated and contradictory motives. So far we've not documented any case of poachers using drones to target wildlife.

6.7.1   DISCUSSION QUESTIONS

1. Should drones be used by non-governmental actors to spy on criminal organizations?

2. Who should have access to this kind of pro-social surveillance technology?

3. Who should regulate this kind of pro-social surveillance technology?

4. What should authorities have done?

5. Does this usage match on our ethical guidelines (Subsidiary, Physical and Material Security, Do No Harm, Newsworthiness, Privacy, Data Protection)?

6.8   CASE 7—HOW HARD CAN IT BE? SECURING FAA PERMISSION TO FLY

One of the first items on our class preparation to-do list was "Secure FAA permission to fly our drones." It seemed like a simple task, especially since Gordon had done all the right things: completed the FAA training and received his part 107 license, filled out all the paperwork, and submitted the FAA waiver request well in advance of the recommended 90-day deadline. As the start of the semester approached, we checked every item off our list, one by one, until the only thing left was securing FAA permission to fly our drones. The semester started, and still we'd heard nothing from the FAA. A *year* after submitting our request we received a response from the FAA: "Unfortunately, your request was not processed due to overwhelming requests in the system. Per your request, you are an educator at a public institution and require no waiver as you are operating for instructional purposes with students. Your request expired on June 1, 2018 and will be canceled" (Personal email to Gordon D. Hoople from the FAA, June 8, 2018). While it is true we are at an educational institution, our university is located in restricted airspace near San Diego International Airport. We absolutely need a waiver to fly on our college campus. Luckily for us we discovered a local hobbyist field near our campus where, for many years, the FAA had allowed model aircraft to fly. After some research we discovered that this approval had also been extended to include drones—a simple solution to our problem.

The FAA process was designed in order to prevent rogue flights by setting out basic policies noted earlier (under 400 ft, within line of site, further than five miles from an airport, not over a crowd, and so forth). These efforts were intended to make life easier for hobbyists while also setting some standards for commercial users. Unfortunately, the implementation process as we experienced it was extremely flawed. Even though Gordon had followed all of the federal instructions, and was one of the first to take the Part 107 certification exam, we could not make it through the bureaucracy.

The FAA process was designed in order to prevent rogue flights by setting out basic policies noted earlier (under 400', within line of site, further than five miles from an airport, not over a crowd, and so forth). These efforts were intended to make life easier for hobbyists while also setting some standards for commercial users. Unfortunately, the implementation process as we experienced it was extremely flawed. Even though Gordon had followed all of the federal instructions, and was one of the first to take the Part 107 certification exam, we could not make it through the bureaucracy.

In the time that's passed since we applied, we often see other folks flying their drones; at the park where Gordon takes his son, over the field where Austin's kids play soccer. Sometimes these drones are flown over people, but for the most part they appear to be operated by responsible pilots. Nevertheless, nearly all of these pilots are breaking the law. The fact is that the place we live, San Diego County, is chock-o-block full of airports. The FAA says one cannot fly within five miles of an airport. With nearly twenty airports in the county, there is virtually no location that is more than five miles from at least one airport. That's why Gordon's certification for permission to fly was so important.

Figure 6.9: Securing permission to fly a drone legally can be a sometimes be a tricky process. iStock image.

As we were writing this book the U.S. Department of Transportation designated San Diego as one of ten cities for testing drone technology. This program aims to examine how drones might be integrated into the fabric of our city. Already we have seen police and firefighters begin to integrate this technology into their workflows. Exciting times! So long as you can get permission.

6.8.1   DISCUSSION QUESTIONS

1. Do you think we should fly our drones on campus without FAA permission?

2. Should the Federal government be in the business of regulating small drone use?

3. How should Federal regulations respond to fast-paced innovation?

4. Since these early days of drone regulations numerous technological solutions have emerged to help simplify the waiver process. Do some research on these methods. Do they appear to be working?

6.9 CONCLUSION

Across these cases we find one constant theme: social, political, scientific, and regulatory efforts lag behind technological innovation. As a result, heterogeneous teams work to deploy new technology for the greater good necessarily face critical decisions about how to proceed. We have chosen cases in which we believe reasonable people can disagree. This chapter is thus designed to spur debate, especially within teams working in this unsettled space.

Several lessons are immediately apparent. First, reasonable people can disagree about (a) what the rules should be and (b) how to apply them. We ourselves don't necessarily agree on reasonable settlements, and don't expect teams of students working across personal and disciplinary differences to all land on the same page. That's exactly the kind of thing sociotechnical thinking helps us anticipate—technologies, whether simple or complex, are the result of many assumptions and leave the creator's control when they enter the real world of use, re-use, remixing, hacking, and so forth. Second, we would like to argue that new technologies do not require new ethics. Existing laws around privacy are probably sufficient, whether the peeping tom is outside one's kitchen window with a camera on their phone, or outside one's penthouse loft with a camera on a drone. Here, too, reasonable people will disagree, and from those conversations societies will arrive at new rules and norms—that is to say, new laws and new ideas of what's okay. In other words, it is likely that a value for privacy persists, but *what* exactly we think is private is likely to change, and how we ensure that privacy is protected is likely to evolve. In closing, we would like to draw attention to a third and final lesson. Readers who hoped for more cases that direct attention to the use of drones for malevolent purposes—creepy spying and delivering drugs to prisons, for example—will be directed back to Figure 6.2. The smallest non-violent uses Austin's team was able to measure were the uses of drones for breaking the law and for spying. We didn't talk much about it because not many people are doing it, at least relative to all the other pro-social uses we found in the data. That finding is one of the reasons we've selected drones as a case study for a book on sociotechnical thinking. We have every reason to believe it can be used for both good and bad, and some confidence that the good will outweigh the bad, all other things considered.

CHAPTER 7

Conclusion

7.1 INTRODUCTION

It may be unusual for a conclusion to fall near the center of a book—indeed we suspect some of the most useful content is situated in the Appendix that follows this final chapter. In the pages that proceed we have done our best to be explicit about our experience and in the pages that follow we hope to be useful and pragmatic in the resources we provide. The Appendices that follow contain our full lesson plan, detailed exercises, and all supporting bibliographic resources. Taken together they reflect our pragmatic commitment to a pedagogy that is fundamentally sociotechnical in its orientation. We hope these tools will be adopted within schools of engineering, and can only dream that it makes its way farther afield, and into the hands of social sciences, and perhaps even into the humanities.

7.2 WHAT HAVE WE LEARNED?

Our primary objective in this volume is to introduce to faculty an approach to educating students on the sociotechnical approaches prevalent in the world they will enter after school. We pursued this goal in a way that we hope makes some sense. In Chapter 2, we introduce our respective classroom philosophies, emphasize the value of a sociotechnical classroom, and detail how we actually taught the class. We also cover what we consider to be some of the best practices in education, as related to the major components in this class. In particular, we focus on the importance of: active learning that engages students in their own education; identifying clear and student-oriented learning objectives; and clarifying the relationship between critical pedagogies and our sociotechnical approach.

In Chapters 3 and 4, we introduce the class' three phases: sociotechnical thinking, drone building, and a pro-social drone project. We then proceed systematically through the steps required at each of these three phases. In many ways this chapter serves as a guide for Appendix 2, where were we provide a detailed lesson plan that covers what we do in every minute of every class. We encourage the reader to flip back and forth between these sections, looking in the main text for an overview and moving to the Appendix for the details of a particular activity that might be of particular interest.

Chapter 5 dives into the social and technical facts and implications of drone technology. We're focused here on consumer-level drones used for the greater good. What exactly are drones?

How do we approach the ethical puzzles they create? We present a number of critical tools for engaging debates over how to balance this new technological capacity with possible social benefits. Our goal in these chapters, and across this volume, is to open debate that goes beyond drones, and in this way provides a viable framework for thinking about and working with emerging technologies.

Chapter 6 presents select case studies that explore many of the common issues associated with drones. We believe case studies are a perfect way to identify and discuss the complex relationships between technology and society, which is why we wrote it in such a way that it can be assigned in the classroom. In each case will present a particular use of a drone and pose discussion questions that surface important tensions between technical, political, social, ethical, economic, and justice dimensions. This approach introduces specific puzzles and highlights key questions for discussion.

In Chapter 7 we have two objectives. The first is stylistic, and is to simply smooth the transition between the preceding chapters *about* the class and the following appendix that are the class. The second is more substantive: We'd both like to get something off our chest. Sure, we came to this project out of a commitment to sociotechnical thinking, and stuck with it because we genuinely like working together. But animating this entire affair is a commitment to see education done in a truly different way. We are neither the first nor the most articulate purveyors of these ideas, but in what's left of this chapter we've included our respective manifestos. We hope they start some fresh conversations on our campuses, and perhaps on yours as well.

7.3   A MANIFESTO FOR ENGINEERING EDUCATION

This class came about because the National Science Foundation solicited proposals to "Revolutionize Engineering Departments." Why, exactly, is a revolution needed? A revolution implies radical change—a major step. What is so broken with our current model? For one, engineering educators have done little to adopt the best practices in education that have been developed over the last 30 years (see Chapter 2). Classrooms look nearly the same as they did a generation ago—with a professor at the front of the class, writing notes on a board, while students furiously scribble them into their notebooks. But this is not a problem unique to engineering—much of higher education has failed to adopt these best practices.

The reason engineering truly needs a revolution is that our discipline's culture has to change. While other professional fields like law and medicine have seen substantial increases in participation by women, participation rates in engineering have remained more or less constant [115]. The argument here is not one for affirmative action or diversity for diversity's sake. Research suggests that diverse teams may produce more creative solutions [116]. For example, if there were more women in engineering, it seems likely that recently released smart speakers would show less of a gender bias when interpreting speech. To address society's most complex problems, we need to assemble teams that can truly think outside of the white, male, heteronormative box.

One narrative paints this as a pipeline problem—there simply are not enough qualified candidates! While it is true that more outreach into K–12 is needed to prepare students for engineering, we argue the real problem lies within engineering culture itself. When women and minorities do choose to pursue a career in engineering, they often find it to be an incredibly hostile culture that rejects their meaningful involvement [115]. Worse still, it seems our education system may be actively discouraging students from adopting a sociotechnical mindset [117, 118]. In a seminal study, Erin Cech explored this "culture of disengagement" and found that engineering students' interest in public welfare actually *declines* over the course of their education [119]. If this doesn't call for a revolution, then we don't know what does.

We built and offer this class in the hopes that we could bring together students of Peace and Engineering in order to prototype new technologies, as well as new ways of knowing, that would promote the greater good. To truly have a positive impact on society, we argue it is imperative to have an understanding of the ways in which forces of power and privilege have lead to the systematic oppression of minority viewpoints throughout history. As any educator who has raised these topics in the classroom knows, many students (but particularly engineers) are not exactly thrilled to discuss these issues.

A different version of our class, one we have contemplated but not yet actualized, could take a deeper dive into issues of social justice. We have approached, but not fully integrated, these subjects within the context of our existing course. For example, one student team proposed designing a drone for surveilling Mexican cartel activity. This gave us a chance to engage in a conversation with students around the impact of America's "War on Drugs" and the disproportionate impact it has had on communities of color [120]. When we helped students to see their proposal in this larger context, they quickly abandoned it. While this conversation steered the team away from this particular drone application, we have yet to implement a systematic approach to helping students design drones that promote social justice. Our experience teaching these themes in this and other contexts makes clear that such topics will engender resistance from some students. Baillie, Leydens, Lucena, and Riley, among others, have written extensively on the topics of engineering and social justice [3, 4, 121]. We encourage readers to familiarize themselves with this body of work and consider ways in which they might integrate social justice into their engineering courses.

7.4   A MANIFESTO FOR PUBLIC POLICY EDUCATION

Academia is a bit of a bummer lately. Tuition is sky high and public funding for education is in the dumps. Our disciplinary standards are narrowing at the very moment borders between major problem and solution sets in the outside world are disappearing. The university sometimes finds itself in the exquisitely unenviable position of gatekeeper for a graveyard, preparing new people with rusty tools. This is a shame, since universities are one of the only places on earth that one can find people with experience on such a wide range of knowledge and skills within a stone's throw from one another.

The class described in these pages was designed in order to draw on this rich potential. Our individual experiences in our respective fields suggest this is where more education should head. Austin, for example, came of age in an era when graduates from schools of public policy and international relations could reasonably anticipate employment within the public sector (within a government, for example) or the nonprofit sector (for a health advocacy group, for example). The last two decades have upended these expectations, as governments outsource key aspects of their work and as nonprofits compete with businesses and optimistic technology firms for talented folks who want to make a difference. One can finish a degree in public policy and get a job working for the Gates Foundation, a private philanthropy outfit that drives health policy and privately sponsor health breakthroughs through the leveraging of its significant funding capacity. But you're likely to be competing for the position with folks who have MBAs, or data mining degrees, or experience in marketing and advertising.

At the same time, many social media firms seem to believe they are facilitating more public connections than any nonprofit champion of civil society could ever dream of. Our students leave the academy and enter a world of tech startups, academic startups, and sovereign startups. They enter this world at the same time the established firmament of international non-governmental organizations is undergoing rapid transformation, and it is not clear that the jobs of 25 years ago will be available to the change-oriented students graduating from a university in the Global North. For example, some reconfigurations are needed to reflect the fact that global inequality is now a bigger problem than global poverty, and that obesity affects more people than does starvation [122, 123].

One implication is that countries that had been net recipients of international aid resources and expertise may now have the educational and financial infrastructure required to envision their own solutions, staff their own initiatives, manage and audit their own books, and in a manner of speaking, be their own change. Schools of global affairs and public policy then face the puzzles of what they are preparing their graduates for, if not a world of Global South changemaking.

Perhaps one possible response lies closer to home. Addressing environmental racism like Flint Michigan's water crisis requires teams of engineers and social scientists. The challenges of addressing racism, sexism, and inequality at home and abroad should keep people trained in thinking sociotechnically busy for a generation. Training people in sociotechnical thinking will put the world on a faster track toward eradicating racism, sexism, and inequality. The status quo has already demonstrated the risks of siloed thinking. The task ahead is to create new programs with greater relevance leveraged alongside our colleagues across campus, across town, and around the world. New ideas and new action are needed. We hope this book contributes to such efforts.

APPENDIX 1

Syllabus

Drones for Good:
Sociotechnical Thinking Across Disciplinary Boundaries

COURSE DESCRIPTION

In our class students from the School of Engineering and the School of Peace Studies come together to design a drone that will have a positive impact on society. You'll have the opportunity to work on an interdisciplinary team in a semester long project-based course. The class is rooted in the social sciences and starts with an investigation of what it means to be an engineer or a peace builder. This is followed by the engineering challenge of building a drone. Underlying the entire semester is a project where you will develop your entrepreneurial skills: design a drone for positive social impact based on an unmet social need.

LEARNING OUTCOMES

At the end of this course you should be able to do the following.

1. Define sociotechnical dualism and explain to a peer why they should approach problem solving from a sociotechnical perspective.

2. Debate issues related to social change; social innovation; power, politics, and inequality; and emerging technologies.

3. Assemble, test, and fly a quadcopter using off the shelf components and open source software.

4. Explain to a friend the major ethical concerns related to drones.

5. Describe the key regulatory guidelines related to drones.

6. Design a drone that can have a positive impact on society (with a critical understanding of the ways drones can negatively impact society).

7. Collaborate on an interdisciplinary team to solve problems and communicate ideas in a compelling way.

8. Acquire new knowledge to address problems outside of your comfort zone.

GRADING PHILOSOPHY

We care deeply about how much you learn in our class. If, in 15 weeks, you leave our class without having discovered something new about both the world and yourself then we have failed you. One way we signal what we want you to learn is through learning objectives. We provide them at both the high level (see above for the course objectives) and the topic level (included on most course materials). We use these to signal to you what we think is important in terms of the material we will cover together. These tell you what we hope you will learn.

We actually care very little about the grade you earn in this class. When you leave here we will not think of you as an A student or a C student. We will think about how much you learned. This is why we want you to focus more on learning and less on grades. We encourage you to read Professor Adam Grant's Op-Ed about grades, because we certainly have! Two quotes for you to consider:

1. Academic excellence is not a strong predictor of career excellence. Across industries, research shows that the correlation between grades and job performance is modest in the first year after college and trivial within a handful of years. For example, at Google, once employees are two or three years out of college, their grades have no bearing on their performance. (Of course, it must be said that if you got D's, you probably didn't end up at Google.)

2. If your goal is to graduate without a blemish on your transcript, you end up taking easier classes and staying within your comfort zone. If you're willing to tolerate the occasional B, you can learn to program in Python while struggling to decipher "Finnegans Wake." You gain experience coping with failures and setbacks, which builds resilience.

All this being said, we do have to assign grades at the end of the semester. We have designed this course so that every student has the opportunity to earn an A. There are no curves and there is no competition between your classmates to earn top marks. The next page explains how we will grade this course. Please come ask us about this if anything is unclear!

COURSE GRADE BREAKDOWN

It is likely this course is graded differently than your other classes. There are two kinds of work in this class: *graded work* and *pass-fail* work. Your grade will be based on a combination of these two components as shown below. The expectation for both graded assignments and pass-fail assignments must be met to achieve the associated course grade. These are not averaged. Performance in one category is in no way a substitute for performance in the other.

Course Grade	Graded Assignments	Pass–Fail Assignments
A	A average	At most 1 missed assignment
B	B average	At most 2 missed assignments
C	C average	At most 3 missed assignments
D	D average	At most 4 missed assignments
F	F average	5 or more missed assignments

For example, if you get an A on your graded assignments, but miss 2 pass-fail assignments, you will get a B in the class. The major takeaway here: your pass-fail assignments are just as important as the graded work, make sure to complete them.

Pass-Fail Assignments

These include homework, reading reflections, final reflection and other "low-stakes" assignments. The reason these are pass-fail is that we want you to have a chance to practice without fear of failing. To earn a passing grade, you need to convince us that you put in a reasonable amount of effort towards the assignment. You don't necessarily need to get the "right" answers (as you'll quickly see there aren't really any "right" answers in this class).

Graded Assignments

The table below shows the major graded assignments and their weighting for the course.

Weight	Graded Assignments
30%	Mid-term essay
20%	Course engagement
50%	Team project (with multiple deliverables throughout the semester)*
*Not all teammembers will receive the same grade. Individual grades on team projects will be determined by faculty based on peer evaluation and in-class observation.	

A Note on Course Engagement

You are one of the best things about this class. When you're gone, people notice, and the class isn't as good. When you are quiet, and don't say what you think or share what you see, then this class isn't as good. It's not as good for us as instructors, and it's not as good for everyone else. For this reason, you will be expected to actively engage with this material outside the classroom, and to demonstrate this engagement through your involvement in the class. Your verbal input is the best indicator of this involvement. If you simply cannot speak up in class, see us in our offices and we will sort something out. Since irregular attendance will disrupt our learning community, absences will affect your grade. Exceptions will only be made *in advance of the class*.

SCHEDULE

Week	Topic	Major Assignments Due
1	Introduction to Drones for Good	
2	How Does Social Change Happen?	
3	What is Social Innovation? (Teams Announced)	
4	What about Power, Politics, Inequality?	Team Drone Idea Due
5	What about New Technologies?	Drone Proposal Infographic—Draft
6	Lab Build: Drones [Build Day 1/2]	Mid-Term Essay Due
7	Lab Build: Drones [Build Day 2/2]	Drone Proposal Infographic—Final
8	Entrepreneurship 101	Case Study, practice flying
9	Presentations, Flight Competition	Team Workplan Presentation, Flying Drone
10	Working day—MVP Research and Development	Social Model Canvas Due
11	Working Day—Presentation Tips	Message Map
Fall or Spring Holiday		
12	Working Day + Pitch Practicing & Feedback	Draft Pitch with MVP Update
13	Working Day + Pitch Feedback	
14	Final Pitch + Course Wrap Up	Final Pitches + MVP Due
15	Final Project Reflection—Due during Exams Week	

COURSE ASSIGNMENTS

All assignments are due at the start of class.

Week 2—How Does Social Change Happen?

Reading and Reflection
Read the articles below and answer the prompt: What is the point of your degree program? What from the reading (if anything) connects or contradicts your answer? Responses should be 300–600 words (1/2 page to 1 page).

- D. M. Riley. "Mindsets in Engineering," in *Engineering and Social Justice*, Morgan & Claypool Publishers, 2008, pp. 33–47 [124].

- J. R. Brown. "Unplugging the GPS," in *Vocation Across the Academy*, D. Cunningham, Ed. New York:Oxford University Press, 2017, pp. 204–224 [125].

- D. Green. *How Change Happens*. New York: Oxford University Press, 2016. Chapter 1 [126] (free download here).

Flipgrid—Individual Drone Pitch

This week in class we gave you some background information about drones. A key project in this class is to create a drone that can have a positive social impact. In this assignment, you need to come up with an idea of how to use a drone to have a positive social impact on society. This idea should be something you would want to work on for the rest of the semester. Once you have identified an idea, you will create a 30-se pitch to convince your classmates to join your team. You can record and upload this pitch using the Flipgrid platform (Flipgrid.com or in your app store) These pitches should simply be a recording of you speaking directly to the camera; no video editing, no text, no GIFs. Your pitch should clearly articulate the problem you are trying to solve, explain how your drone will have a positive social impact, and persuade your classmates to work with you. We will watch these videos together in class.

Week 3—What is Social Innovation?

Team Preference Form

Please fill out the Google form by Sunday 9PM so we have time to form team before class.

Reading and Reflection

Read the articles below and upload and respond to the prompt: *What's the overlap, if any, between Social Justice and Social Innovation? Are they compatible? Incompatible?* Responses should be 300–600 words (1/2 page to 1 page).

- Schwab Foundation. *What is Social Entrepreneurship?* Online article here.

- D. Riley. "What do we mean by social Justice?," *Soc. Learn. Technol. Innov.*, pp. 47–93, 2005. [127]

- A. Choi-Fitzpatrick. *The Good Drone: How Social Movements Democratize Surveillance*. Cambridge, MA: MIT Press, 2020. Chapter 1. [10]

Brainstorm Drone Ideas

Brainstorm three ideas for how drones could have a positive impact on society. Turn in a one sentence description for each idea and bring these ideas to class to discuss with your teammates.

Week 4—What about Power, Politics, Inequality?

Reading and Reflection

Read the articles below and respond to the prompt: *What are the key arguments made by each of the authors this week? Do you agree or disagree?* (~1 paragraph per reading/video) Responses should be 300–600 words (1/2 page to 1 page).

- Morozov, Evengy. (2013) Video Interview with The Economist here (5 min).

- A. Choi-Fitzpatrick. *The Good Drone: How Social Movements Democratize Surveillance.* Cambridge, MA: MIT Press, 2020. Chapter 2 [10].

- D. Nieusma and E. Blue, "Engineering and War," *Int. J. Eng. Soc. Justice, Peace,* 1(1), pp. 50–62, 2012 [128].

- D. Green. *How change happens.* New York:Oxford University Press, 2016. Chapter 2 [126] (free download here).

Choose Drone Topics [Team]
We went through a rapid ideation process in class last week. We expect you to continue this discussion between last week and today. Bring to class your team's three–to five best ideas and identify what problem your team is interested in most solving with a drone. Come prepared to discuss your idea in detail with the instructors in class.

Week 5—What about New Technologies?

Reading and Reflection
Read the articles below respond to the prompt: *Think about our empathy essay in relation to the video you watched. Do you think empathy really matters?* (It's totally fine to call us crazy!). Responses should be 300–600 words (1/2 page to 1 page).

- Watch one of the following on Netflix:

  ◦ *Her* [120 min]

  ◦ *Black Mirror: White Christmas* [74 min]

  ◦ *Black Mirror: Hated in the Nation* [90 min]

- G. D. Hoople and A. Choi-Fitzpatrick. "Engineering Empathy: A Multidisciplinary Approach Combining Engineering, Peace Studies, and Drones," *ASEE Annu. Conf. Proc.*, 2017 [13].

- M. Ganz, T. Kay, and J. Spicer. "Social Enterprise Is Not Social Change," *Stanford Soc. Innov. Rev.*, 16(2), pp. 59–60, 2018 [129].

- A. Choi-Fitzpatrick. *The Good Drone: How Social Movements Democratize Surveillance.* Cambridge, MA: MIT Press, 2020. Chapter 3 [10].

Drone Proposal Infographic + Summary
Draft Due: Week 5
Final Due: Week 7

Aligning with the requirements for the USD Social Innovation Challenge, in this assignment you will create an executive summary and a visual representation of your idea for how to use a drone for the greater good. *Note that in this class we are flipping the engineering design process and letting the solution (drones) drive the problem identification. This is because (1) we want you to get experience thinking about how drones interact with society and (2) because the semester is short and we don't have much time!*

As a result, the solution landscape is constrained such that you must use a drone. A part of the challenge is finding an appropriate problem and rejecting interesting problems that are better solved with alternative technologies.

Submissions should address three important question areas and be focused on a single, focused issue.

Problem Landscape	Solutions Landscape	Your Solution (Drone)
What is the issue you are looking to understand? What are the political, economic, social, technological, environmental, and legal hurdles preserving the status quo? Who is affected by it? What is the size and scope of the issue? What is the relationship of this problem to other areas of concern or opportunity?	Who is already trying to solve this problem? What are they doing? What efforts have been tried or are being tried? What has worked, what hasn't? Are any of these efforts linked to one another? What networks and resources exist?	Why are drones the right technology to address this problem? How do they improve on existing solutions? Are they more just, effective, sustainable, scalable, affordable, etc.? We suggest you include information from a SWOT analysis.

You should present this information in two ways:

- *Executive Summary*: 600 words (~200 prob/~200 solutions/~200 drone)

- *PDF Infographic*: We suggest you start with a template on Canva. Additional tips available here.

Week 6—Lab Build—1/2

Midcourse Feedback Survey

This is anonymous! We'll use this feedback to tweak the rest of the semester.

Midterm Essay

This midterm essay is designed as a reflective exercise to help you connect what we have been doing in class more broadly with your education at USD. Recall on the first day of class we talked about disciplinary lenses (Engineer/Kroc School). This essay challenges you to dive deeper on this topic.

Your essay should begin by clearly articulating the disciplinary lens that you bring to this class. What are strengths of this approach of viewing the world? What are some weaknesses with this approach?

Next, considering the readings, videos, and class discussions from the course so far, identify *three key themes* that have challenged how you think about your discipline. Briefly explain these themes (with reference to the appropriate class material) and explain how they have impacted your thinking. You should feel free to agree or disagree with any of this material. The objective here is to demonstrate that you have engaged with the course material in a meaningful way. We are not looking for an essay describing how you completely agree with all of the material we assigned. Quite the contrary, we want you to wrestle with these ideas and see if they hold water.

Your essay should be written in the first person and have a clear logical flow. It should be 3–5 pages, single-spaced, and have size 12 Times New Roman font.

Week 7—Lab Build—2/2

Reading (no reflection)

- I. Frazier. "The Trippy, High-Speed World of Drone Racing," *The New Yorker*, 2018. (here) [130].

- H. Babinsky. "How Do Wings Work?," *Phys. Educ.*, 38(6), pp. 497–503, 2003 [131].

Drone Proposal Infographic—Final DUE [See Week 5 for Assignment]

Week 8—Entrepreneurship 101

Reading

- Skim

  ◦ Burkett, Ingrid. (No date) "Using the Business Model Canvas for Social Enterprise Design" (here).

- Read/Watch

  ◦ N. Hughes and S. Lonie. "M-PESA: Mobile Money for the 'Unbanked' Turning Cellphones into 24-Hour Tellers in Kenya," *Innov. Technol. Governance, Glob.*, 2(1–2,) pp. 63–81, 2007 [132].

  ◦ Video 1 from Safaricom

  ◦ Video 2 from USAID

- Also, be sure to practice flying for next week's flight competition!

Weeks 9-14—Drones for Good Final Project

Your assignment for the rest of the semester is to refine and validate your idea for a drone that will have a pro-social impact. While the ideas you presented in your Infographics are a good start, they still require many key details. You will spend the rest of the semester working to flesh out these ideas. Your deliverables at the end of the course are twofold:

- a minimum viable product for your drone.

- an 8-min pitch (presentation) to convince the relevant stakeholders to adopt your idea.

In addition to these final deliverables, we have several interim deliverables and drafts to help you keep on track. The schedule below should guide you as you work on this project.

Week	Deadline
9	Team Workplan Presentation
10	Social Model Canvas
11	Message Map
	Thanksgiving—No Class
12	Draft Pitch, 75% Complete MVP
13	(not deliverable)
14	Pitches + MVP Due

Final Deliverables
Minimum Viable Products (M.V.P)

- A Minimum Viable Product is a version of a new product that allows a team to collect the maximum amount of validated learning about the project with the least effort. Recall from the video the example presented by Steve Blank. The team building an autonomous lawnmower kludged together a cart they dragged through a field to show they could in fact identify weeds using their computer vision algorithm.

- Your MVP should be something that helps you to learn whether your idea for a drone is feasible. It should help you address the following questions:

    ○ Is a drone the right solution for the problem?

    ○ Does the creation of this drone serve the public interest?

    ○ How do you avoid doing harm?

    ○ What are the key issues for privacy and data security?

- You should consider issues from multiple angles, including Political, Economic, Social, Technological, Environmental, and Legal and expect questions on the same.

Pitch

- Your pitch must be a persuasive presentation and make an "ask" from the stakeholder you are targeting that will move you *from* your MVP to the next stage of your project.

  ◦ For example—$2,000 from the Social Innovation Challenge to build your next prototype

- Your pitch should cover each of items covered in the infographic:

  ◦ Problem Landscape

  ◦ Solution Landscape

  ◦ Your Solution

- The pitch will leverage knowledge gained in each of the course deliverables, but will synthesize them in a fresh and compelling way.

- Each member of your team must speak during the presentation.

- Drawing on the approaches developed this semester, your team will present your idea to the class and a panel of experts we will invite to the final day of class.

- Pitches will be 8 min long with 4 min for questions and answers.

- ALL PRESENTATIONS must be loaded (1) before the start of class, (2) in Google Slides or PowerPoint format, and (3) into our class folder on Google Drive.

Interim Deliverables

Team Workplan Presentation—Week 9

Your presentation will specify what your team will be doing for the rest of the semester. This is the process whereby your team will generate its MVP and Pitch. The presentation should be 7 min long (maximum, shorter is okay) and answer the following questions as efficiently as possible (e.g., ~7 slides):

- Title Slide—Team name, team members, class name, date.

- Problem Landscape—What's the problem you're using a drone to solve?

- Solution—What does your drone do to solve this problem?

- MVP—What is your MVP and how will it take your Solution to the next level?

- Major Tasks—What needs to be done to finish your project?

- Roles and Responsibilities—Who will do what, and when?

- Resources—What else is needed in order to complete your project?

ALL PRESENTATIONS must be loaded (1) before the start of class, (2) in Google Slides or PowerPoint format, and (3) into this folder on Google Drive.

Social Model Canvas—Week 10

- Please use this free online tool to build your Social Model Canvas.

- Here is an example for a company called KIVA, note it is not exactly the right format and each sticky should have more detail, but it gives you some idea.

- Refer to these two documents to help you think about what to put in your social model canvas:

  - Social Model Canvas (Pages 8–10 particularly helpful)

  - Business Model Canvas (More in depth if needed)

- Recall that the social model canvas should have two statements for each section:

  - Commerce Statements (related to the business you are proposing)

  - Impact Statements (related to how you will have a social impact)

  - Use two colors! One for commerce, one for impact!

- Please email us a link to access your social model canvas.

Message Map—Week 11

- A message map is central to your pitch. It is here that you will refine how the problem landscape, solution landscape, your solution fit together in a cohesive story.

- See this Forbes article for more information. Essentially:

  - Step One. Create a Twitter-friendly headline.

  - Step Two. Support the headline with roughly three key benefits.

  - Step Three. Reinforce the three benefits with stories, statistics, and examples.

- This can be a simple wire-frame mock-up on a white-board or blank paper

Draft Pitch + 75% Complete MVP—Week 12

- Your draft pitch should contain all of the key elements that will be a part of your final pitch, however it does not yet need to be polished.

- We will have each team do a dry run of their presentation for the instructors and one other team. At a minimum you should have a cohesive storyline and rough slides that support the overall message. The more developed your presentation, the better feedback you will receive.

- At this point in class your MVP should also be 75% complete

Week 15—Final Project Reflections

Using this template, complete this assignment individually, without pressure from your teammates. Your responses will be confidential to the instructor, please be candid. This assignment will be graded only for the inclusion of meaningful and thoughtful content. The specifics of what you say and the scores you award will not impact your assignment grade.

Part I: Team Dynamics Reflection

Please write two to three paragraphs reflecting on the team performance as a whole. The team evaluation should address, but need not be limited to, the following questions:

- How did the team function overall? What did the team do well as a whole? What things could the team have done better on?

- How did members of the team communicate? Where were there miscommunications? What caused them?

- How did the range in expertise impact the work of the team (e.g., some students more comfortable in technical fields while others more at home on social issues)?

Part II: Final Project Reflection

Please write two to three paragraphs reflecting on the project as a whole. Your reflection should address, but need not be limited to, the following questions:

- How was the topic for the final project chosen?

- Did you feel included in that decision?

- To what extent does the final product reflect your values?

- Do you feel excited about the final product?

- Do you have concerns about the final product? If so, what?

- Do you think the entrepreneurial skills you learned in class will be valuable?

Part III: Peer Evaluation, Numeric

For each member of the team, including yourself, assign a number between 1–5. Really use the full range of numbers from 1–5. If you simply "award" scores of 4 and 5, then team members who have worked especially hard or provided extraordinary leadership will go unrecognized, as will those at the other end of the scale.

1 – Very Poor (Not many people should be getting this score)

2 – Below Average

3 – Average

4 – Above Average

5 – Outstanding (Not many people should be getting this score)

Team Members: Name → Rating Categories	(1 – Self)	(2)	(3)	(4)
Demonstrated Effectiveness				
Quality of Work				
Leadership				
Commitment to Team and Project				

Demonstrated Effectiveness: Has this person done what is promised? Does she/he contribute to (or detract from) the overall progress of the team? What would happen if this person was missing?

Quality of Work: Is this person's work correct, clear, complete, and relevant? Are analyses, charts, notes, and graphs clear and intelligible? Does this person create effective reports and presentations?

Leadership: Does this person initiate activities, make suggestions, provide focus?

Commitment to Team and Project: Does this person care about the project? Does she/he attend all meetings, arrive promptly, come prepared and ready to work?

Part IV: Peer Evaluation, Written

Please write one paragraphs about each member of your project team, including yourself. The individual evaluations should address the following.

- What were the strengths and areas for improvement for each individual?

- Did this person's performance change over the course semester? If so, how?

- What were the roles that each team member played?

Please be specific and give examples. A vague comment like "Teammate A is great to work with" is not very useful and does not help that teammate understand what to do to improve. A better response would be "Teammate A is open to the ideas of other teammates. For example…"

APPENDIX 2

Lesson Plans

A brief note on scheduling: We have chosen to run this class as once a week in a single 3-hr seminar. These large blocks of time help students engage more deeply in class discussion and in designing and building their drones. We encourage other instructors to experiment with a format that works best in their institutional context.

WEEK 1: INTRODUCTION TO ENGINEERING PEACE WITH DRONES

Class Prep

Assignments Due:

- None

Before class:

- Review roster

Bring to class:

- Syllabus

- Schedule

- Index cards

- Sharpies

- Anticipation essay worksheet

  ○ Prompt: What do you expect to get out of this class?

By the end of this lesson students should be able to do the following.

- Identify the other students in this course.

- Find important information on the syllabus and list the major deliverables for the class.

- Explain something about their world (Engineering or Peace Studies) to someone outside their discipline.

- Describe at least three different applications of drones and any associated ethical concerns with those uses.

Class Content

Class Introduction [30 min]

- This content block includes a round of introductions and an icebreaker.

- Introduction to instructors, the class, a reflection on the importance of empathy, and an introduction to cross-, inter-, intra-, and multi-disciplinary teamwork.

  - Instructor introductions

    - Personal [family/hobby], Background, and Why this class

    - Stuff to signal:

      - This class and collaboration are unusual in university context.

      - This class is an experiment (goal: set expectations).

      - Drones aren't the point, collaboration is ...

      - Faculty are here to learn too (but you're in safe hands).

- Brief introductions around the room (names/major).

- Hand out index cards and ask students to write: Name, Degree Program, Why they chose this class, Anything we should know about you?

  - Collect cards

- Rock Paper Scissor Champion exercise [Icebreaker]

  - In this exercise individuals pair up to play rock-paper-scissors. The loser in each round cheers on the winner by name as the game is played until the entire group has paired off and there is one winner (with a whole class cheering them on). Start with student pairs exchanging names.

  - https://www.icebreakers.ws/large-group/extreme-rock-paper-scissors.html

Disciplinary Identity [30 min]

- This content block focuses on exploring students' disciplinary identity.

- Exploring your identity.

  - Ask students to grab someone from a different discipline.

- Individually, students make a plan for how to answer: [2 min]

    - What are you studying?

    - Why are studying that?

    - What are key classes or skills in your program?

    - What assumptions do you make about the other student category in the class?

- Explain your answers to partner from other program [7–10 min]

- Class Discussion [15 min]

    - Facilitating tip: ensure multiple voices from across and within programs

- Surface and raise common assumptions:

    - Example: Peace School folks care about making and changing social stuff

        - Counterpoint—Austin cares about how change-agents use tools and technology!

    - Example: Engineering School folks care about making and changing physical stuff

        - Counterpoint—Gordon cares about how stuff interfaces with the world people live in!

- Identity shapes the spaces we work within

    - Visual Tour (3 slides): Austin shows pictures of Peace School that everyone's familiar with:

        - Garden of the Sea/Sky: pretty traditional

        - Solar Array and Interactive Art Walls: trying somethings new

    - Actual Tour: Engineers lead other students on tour of engineering spaces

        - Some spaces are outsider-friendly (e.g., easy badge for the Donald's Garage maker space)

        - Some are not (machine shop, electronics labs)

Break [10-15 min]

What are Drones? [60 min]

- This activity is designed to help students explore their preconceptions about drones

- Think Pair Share [with neighbor] [30 min]

    - Individual brainstorm:

- What is a drone, in your opinion?

- How have you seen drones used?

- What are your concerns with drones?

  ○ Prompt students to discuss answers with a neighbor from a different program (preferably not same partner as before)

  ○ Class discussion—what themes emerged?

    - Facilitation tips:

      - Highlight Quadcopters v. UAV

      - Students usually have varying concerns based on what technology they are imagining

      - There are lots of concerns, broadly can be grouped into ethical, safety and security, privacy, etc.

- Drones Overview Presentation [30 min]

  ○ Gordon presents various types of military drones and some histories of each

  ○ Austin presents overview of his book focused on civil applications of small quadcopters

Course Logistics [45 min]

- Flipgrid Drone Use Assignment [10 min]

  ○ Explain that we will be using the Flipgrid platform to quickly capture student pitches.

  ○ Pitches are 30 sec to explain a way to use drones for the public good.

- In-class Essay [15 min]

  ○ Take a minute to free-write your response to the following questions:

    - What do you expect to get out of this class?

    - What are your hopes and fears around this class?

    - What are key classes in your discipline?

    - How do these classes train you to be an Engineer/Other Program?

- About the syllabus [10 min]

  ○ Read the syllabus on your own.

- Turn to your neighbor and share any questions you may have with them.

  - Class asks professors unresolved questions.

- Review assignments for next class! [10 min]

  - Assigned reading and reflection.

  - Thirty-second pitch uploaded to Flipgrid.

  - Set expectations, remind students these are pass fail assignments.

WEEK 2: HOW DOES SOCIAL CHANGE HAPPEN?

Class Prep

Assignments Due:

- Reading and reading reflection

- Flipgrid Videos

Before Class:

- Test Audio/Visual for Flipgrid video display

- Review readings

Bring to class:

- Sticky notes for students

- Paper for note-taking during flip grid

By the end of this lesson students should be able to:

- Give examples of the role of your field/discipline in social change.

- Debate the benefits and harms of technology in social change.

- Identify two elements of a successful pitch.

- Provide actionable feedback about pitches to their peers.

Class Content

Reading Discussion [30 min]

- In this discussion-based exercise students compare and contrast the articles in small groups.

- Break class up into six groups, each in charge of presenting a summary of one article (two groups per article) [10 min]
  - Riley, Mindsets in Engineering.
  - Brown, Unplugging the GPS
  - Green, How Change Happen Chapter 1
- Two groups get together and compare and contrast their two respective readings [10 min]
  - Riley readers chat with Brown readers.
  - Riley readers chat with Green readers.
  - Green readers chat with Brown readers.
- Report back to the entire class on trends, themes, and tensions [10 min]

Debate! What is the Role of Technology in Social Change [60 min]

- In this exercise, always a popular one, we challenge student to debate the declarative statement Technology is having a positive impact on society.
- Class splits into three groups:
  - Judges (n=~3–4) (We take volunteers)
  - Pro team arguing in support of the statement (n=10) (randomly assigned)
  - Con team arguing against the statement (n=10) (randomly assigned)
- Teams: Plan out your side's argument [20 min]
  - Draw on reading as you prepare your arguments
- Judges: Prepare your questions for the two sides:
  - For POSITIVE:
    - For example, What about Killer Robots!!!!
  - For NEGATIVE
    - For example, What about Science, Tech, Medicine!?
- Debate! [20 min]
  - Opening statement from each side [3–5 min]
  - Questions from judges [5 min]

- ○ Teams regroup and prep final statements [10 min]

- ○ Teams deliver final statements [5 min]

- Class Break, while judges make decision

- Judges announce decision after the break

Break [10 min]

Watch Flipgrid Responses in class [50 min]

- In this exercise students get their first practice at pitching an idea. We intentionally left the assignment vague to see how students would handle the activity—it then becomes a useful reference throughout the semester.

- Before showing the videos, lead a discussion on how to give constructive feedback [10 min]

  - ○ Core message: Focus on actionable and concrete comments.

  - ○ Ask students to think of times when they have gotten good/bad feedback.

- Faculty write question prompts on the board—ask students to take notes as we watch pitches.

  - ○ What makes effective pitch?

  - ○ What hurts a pitch?

- Watch Flipgrid videos [approximately 20 min]

  - ○ Everyone takes notes on what works and what doesn't.

  - ○ Everyone notes their favorite things and prepares to explain why.

  - ○ Faculty—pay attention to drone themes raised by students.

- Group Discussion at your table

  - ○ What worked/didn't work?

  - ○ Encourage the group to develop effective pitch criteria .

  - ○ What are the top 1–3 pitches for the effective pitch criteria you've developed?

- Groups report back to class:

  - ○ What works?

  - ○ What doesn't work?

- º What were the favorite pitches?

- º What are common themes folks observed?

- Faculty coordinates summative feedback on what worked and what didn't work *overall* and *across* [all] *presentations* (remembering constructive feedback notes).

  - º Often there are disciplinary trends evidenced by pitches—keep an eye out and surface these. In our class the peace students were very focused on the WHY but had limited understanding of technical details. Engineers had lots of technical expertise, but no real vision for how to make the drone have a positive impact.

  - º Students often forget to say their names.

  - º Some record in noisy environments, while other take care to choose secluded/ beautiful locations.

- Faculty share with the class broad categories of drone use that they heard in the pitches.

  - º Write these on the board and ask students to add/modify/critique.

  - º Indicate these will be on the team preference form, so if there is anything missing students want to work on they should speak up!

  - º Once a list is established, save it so that it can be used to generate team preference form.

Class Conclusion [30 min]

- Course Logistics [5 min]

  - º Tease the Entrepreneurship aspect of course towards end of semester.

  - º Show the schedule on the overhead.

  - º Show major assignments documents to reiterate what is coming.

- Sticky Note Reflection Session [20 min]

  - º Do Sticky Notes for each answer and put up on wall. Students and faculty should group them dynamically. Based on today's class, what is:

    - Something you learned about your discipline.

    - Something you saw from a new perspective.

  - º Faculty review, analyze, and reflect out to class where our overarching themes overlap with things the students have written.

- Assignment for Next Week [5 min]
  - ○ Reading + Reflection prompt
    - "Read the articles below and upload a .doc or .pdf to blackboard that responds to prompt. Responses should be 300–600 words (1/2 page to 1 page).
  - ○ Team preference form (due in a few days!)
  - ○ Come prepared to share three brainstorm ideas you have for using drones for the public good.

WEEK 3: WHAT IS SOCIAL INNOVATION?

Class Prep

Assignments Due:

- Reading and reading reflection
- Brainstorm three ideas you have for using drones for the public good

Before Class:

- Form teams based on team preference form
- Get to class early and put teams on screen
- Prompt students to (1) open their Reflection Essay online and (2) sit with their group at a table.
- Review readings
- Grade and return (pass/fail) previous weeks reflections

Bring to class:

- Sticky Notes
- Markers

By the end of this lesson students should be able to:

- Define the term *social innovation*.
- Articulate a definition of social justice that resonates with your values.
- Describe the basic elements of the engineering design process.

- Brainstorm ideas using a "Yes, And…" approach.

- Apply SWOT analysis to select from among a variety of options.

Class Content

Announce Teams [15 min]

- Put formed teams up on a slide before class begins and ask students to sit with their teams.

- Start class with some team formation icebreakers (choose your favorite).

- Reviewing teaming best practices—we cover stages of team formation and perform some norming activities to help students get on the same page as each other.

Defining Social Entrepreneurship and Social Justice [50 min]

- This activity challenges students to engage with the readings and define for themselves (and then as a team) the nebulous concepts of social entrepreneurship and social justice.

- *Challenge students* to define the terms Social Justice and Social Entrepreneurship *On your own* (quietly). You can create your own or adopt another's definition from the readings, but must be prepared to justify your choice [5 min].

- *As a group* decide on a shared definition of each term [10 min].

- In *pairs of groups* report out/share your definitions/discuss similarities and differences [25 min].

- Faculty put up their definitions, but emphasize there is not one correct answer [10 min]

  ◦ We emphasized normative aspects of definitions. Social Justice tends to be inherently normative in its definitions while social innovation doesn't have to be.

  ◦ Field questions and/or a brief discussion on the contradictions, gaps, and/or themes that emerged from this exercise.

How We Generate Ideas [20 min]

- This activity is aimed at helping students generate a set of possible topics for the semester long project of designing a drone for good.

- Start with very brief explanation of the engineering design process—this is geared for students who don't have a background in the field [<5 min].

    - Emphasize we are short circuiting a little bit by forcing the use of drones into the equation. Not good design practice, since you want to keep your options open, but this is a good pedagogical exercise since it forces us to focus for action.

- Group Activity: "Yes But vs. Yes And"

    - Goal: Get teams to see the value of building on others' ideas rather than shooting them down, thereby generating more possibilities.

    - Note: This is based on Exercise 3 in the Stanford D-School *Stoke Deck*.

    - Punchline: *Yes, And* creates openings, *Yes, But* creates closures.

    - Description of the activity:

        - Part One: In their teams, one student seeds an idea (e.g., *Let's have a party*) and the rest of the team contributes their negative "Yes, But." As in: *Let's have a party. Yes, But it could be raining.*

        - Part Two: In their teams, the same student seeds an idea (e.g., *Let's have a party*) and the rest of the team contributes their positive "Yes, And." As in *Let's have a party. Yes, And we could invite the whole neighborhood.*

    - Most important: Debrief the activity so students see why we want to build on the ideas of others.

Break [10 min]

Post It Brainstorming on Drones for Good [40 min]

- In this activity we apply the lessons from the Yes, And approach to generate ideas about using drones for the greater good. The goal is to quickly generate lots of ideas, not necessarily good ones.

    - Make sure each team has Sticky Notes and markers

- Description of the activity:

    - In their teams, and preferably working on a whiteboard or wall, students play the *Yes, And* game with "How might we design drones for good?" as the prompt.

    - After 3–5 min, add a challenging constraint (e.g., must cost less than $100, must involve liquids, etc.).

- Continue adding several rounds of constraints to help students think outside the box. We strongly recommend including several silly constraints such as "all ideas for the next 2 min must include bananas." These help students stretch their imaginations.

- Toward the end of the activity ask students to review all ideas and group them by themes.

- Activity wrap up:

  - Emphasize this is just the start of the idea generation process, we expect teams to continue generating more ideas before the next class.

  - Group reflection on exercise—what worked, what didn't.

SWOT Analysis and How to assess ideas [30 min]

- This activity teaches students one method for choosing from among potential ideas: SWOT (strengths, weaknesses, opportunities, threats). You could easily swap this out for another idea evaluation criteria approach.

- Faculty introduce SWOT analysis as a tool for evaluating ideas. We based our introduction on the following resources:

  - https://formswift.com/business-plan#swotanalysis

  - http://ctb.ku.edu/en/table-of-contents/assessment/assessing-community-needs-and-resources/swot-analysis/main

  - https://www.mindtools.com/pages/article/newTMC_05.htm

- Student SWOT Analysis Exercise

  - The *Yes, And* brainstorming activity generated lots of ideas. How to separate out the good from the bad?

  - As a team, apply SWOT approach to the ideas that emerged from the *Yes, And* brainstorming activity.

Class Conclusion [15 min]

- Check in on reading and reflections. How's that going?

  - Encourage Peace Studies students to share reading best practices to help deal with the volume of reading.

- Reminders for next week:

- ○ Individual: Reading and Reflection

- ○ Group: Generate more ideas and down select to 1–3 projects they want to work on.

WEEK 4: WHAT ABOUT POWER, POLITICS, INEQUALITY?

Class Prep

Assignments Due:

- Reading and reading reflection

Before Class:

- Review readings

- Grade and return (pass/fail) previous weeks reflections

Bring to class:

- Sticky flipchart paper (for infographic prototypes, if you don't have white boards)

- Sticky Notes

By the end of this lesson students should be able to:

- Recognize and explain the concept of technological solutionism.

- Describe the challenge of unintended consequences in innovation.

- Generate rapid prototypes (through rough sketching) of infographics and posters.

- Provide actionable feedback to their peers.

Class Content

Group Discussion: Power, Politics, Inequality [55 min]

- This activity, about power, politics, and inequality, is structured in such a way that students experience the impact of power structures as a part of the discussion process. It is often our most powerful discussion activity.

- Description of Activity:

  - ○ Set up the room with (1) a core set of chairs or tables in the center and (2) a peripheral set of chairs arranged around the core.

    ° Conduct two consecutive fishbowl-style discussions, the first with Engineering School students discussing Peace School reading followed by Peace School students discussing the Engineering School reading.

    ° Discussion 1 [10–15 min]:

- Engineers in the center/hot seat.

- Peace School students at the periphery, as onlookers.

- Faculty say that every student must talk at least once, but that we won't be involved/facilitate the discussion in any way after putting up the prompt.

- Discussing prompt: *What is power (according to Duncan Green) and how is it used?*

- Discussion:

    - For the Peace School student onlookers: *How did that go? How did it feel to watch? Anything you really wanted to say?*

    - For Engineering students on the hotseat: *How did that feel being in the middle?*

    ° Discussion 2 [10–15 min]:

- Peace School students in the center/hot seat.

- Engineering School students at the periphery, as onlookers.

- Discussing prompt: *Reflecting on Nieusma and Blue's article on engineering and war: Does it matter where engineering came from? Did this reading change your perception of engineering?*

- Discussion:

    - For the Engineering School student onlookers: *How did that go? How did it feel to watch? Anything you really wanted to say?*

    - For Peace School students on the hotseat: *How did that go?*

Review Possibilities and Choose Projects [60 min]

- Launch this activity by explaining to students the tensions between technological feasibility and positive social impact. Some things that are easy to do don't have much impact, and some things that would have the most impact are perhaps impossible.

- The team has brainstormed ideas (via *Yes, And*) and thought critically about them (via SWOT analysis). Now's the time to make some tough choices.

- Prompt students: as a team, graph your top three ideas about drones for positive social impact. Let:

    - X = Technological Feasibility

    - Y = Positive Social Impact

    - With this frame in mind discuss in your team what idea you want to work on?

- Faculty meet with teams one-on-one to review team's ideas.

    - While teams are working, circulate to discuss their ideas, doing your best to weed out ideas deemed to have limited likelihood of success.

    - Goal: Help team arrive at one solid idea.

    - Prompt: Ask students to think back to their Flipgrid pitches—how would they pitch this idea, and to whom? Ask students to start developing a plan for pitching this idea (PROTIP: they'll use that in the next exercise).

Break [10 min]

Rapid prototyping of infographic [50 min]

- This activity, where students quickly build an infographic and get feedback, is geared toward having students further develop the idea they have chosen in the previous activity.

- Faculty explain to students we'll be building an infographic for their idea, but before we do that let's talk more about them! [2 min]

    - Everyone's seen one, but what makes them good?

- Student review example infographics [10 min]

    - Students open laptop per table, follow the example infographic links distributed to class. We recommend picking a wide range, we used the ones below:

        - https://goo.gl/M6YAvt

        - https://goo.gl/dRcBJi

        - https://goo.gl/8NKo5G

    - Discuss with your team: What makes them effective?

    - Discuss with your team: What are common pitfalls?

- Faculty-facilitated conversation: What are best and worst practices in infographics? bring together learning from each team [5 min]

- Details on our Infographic assignment [3 min]

    ° Tell students about their assignment: Infographic + Executive Summary.

    ° For details, see "Course Assignments" in the syllabus.

- Introduce rapid prototyping [5 min]

    ° Explain we're using "rapid prototyping" process to quickly mock-up infographic.

    ° Prompt students to sketch out their wireframe infographic on paper or whiteboard.

    ° Post it up when done and prepare for feedback.

- Infographic Gallery Walk [10 min]

    ° Reinforce for students that this activity is about both giving positive and constructive feedback. Remind them that we are looking for actionable (as opposed to vague) comments.

    ° One team member stays with your infographic to pitch your idea to visitors, giving more practice pitching.

    ° Other team members rotate around and provide feedback on Sticky Notes.

        - Something you like about the infographic.

        - Something about the infographic you'd change.

        - Is the idea viable? If not, why?

    ° Depending on class size—complete 3–5 rotations so teams get multiple different sets of feedback.

- Teams make a plan for next week [10 min]

    ° Debrief on the feedback your team received in the Infographic Feedback Session (and ideas you saw on other infographics you want to incorporate).

    ° Inventory your assets.

    ° Assign key tasks.

    ° Set a meeting time outside of class.

Class Conclusion [5 min]

- Due next week:

    ° Draft Infographic + Executive Summary

    ◦ Reading and reflections

WEEK 5: WHAT ABOUT NEW TECHNOLOGIES?

Class Prep

Assignments Due:

- Reading and reading reflection

- Team's draft proposal

Before Class:

- Review readings

- Grade and return (pass/fail) previous weeks reflections

Bring to class::

- Markers

- Large sticky-note flipboard paper for discussion points

- Reading prompts, printed out one per 8 ½ × 11 page and affixed to the large flipboard paper

- 8 ½ × 11 paper for sketching

By the end of this lesson students should be able to:

- Articulate their personal views about the ways technology is transforming societies and economies.

- Explain to a peer the difference between sympathy and empathy.

- Create a plan of action based on peer feedback.

- Describe some of the challenges and benefits of interdisciplinary teamwork.

Class Content

Reading discussion: Thinking about new technologies [60 min]

- This activity provides a more dynamic way to engage with a range of questions about the readings.

- Distribute one prompt per table on large flip board paper with markers. Prompts:

- ○ What new technology are you most excited about?

- ○ What new technology makes you the most nervous?

- ○ What's best way to balance tech's promise/peril?

- ○ What percentage of a technology's impact should we have thought through before continuing to production (think Morozov's *Unintended Consequences*)?

- ○ Who should help manage trade-offs? Governments, businesses, activists, citizens, consumers?

- ○ What video did you watch, and what message does it provide on power (recall the readings from Duncan Green)?

- Invite teams to discuss the prompt they find at their table.

  - ○ As a team, discuss the question prompt you find at the table.

  - ○ Write down on the flip chart your response(s) to the question. Remind students they don't have to agree!.

  - ○ Spend ~5 min per table.

  - ○ Rotate to the next table, which has a new question. Quickly read what others have written and then continue the discussion.

  - ○ Continue this rotation until each team reaches their original station.

  - ○ Spend 5 min synthesizing the comments on your original prompt.

  - ○ Report out to the class.

Building Empathy: Memorable Moments [15 min]

- This activity is designed to get students to engage with each other in more dynamic ways that push them to develop empathy for each other.

- Start by giving students an example of a memorable moment of your own—a personal story about something memorable.

  - ○ For example, Gordon told this story to our class: The first time I got to install hundreds of sensors on a large aircraft. In particular, I recall pushing on the airplane (which was suspended in a free-free boundary condition) and because of the scale I couldn't tell if the plane was moving or I was moving.

- Have all of the students to think of a memorable moment—draw a sketch of that moment. [5 min]. Note this doesn't have to be their most memorable moment, just something memorable.

- Look at your picture—what is the emotion that it makes you think of—keep that emotion to yourself.

- Give your picture to your neighbor. Based on the picture ask them to write down on the back what emotion they think of.

- Once everyone has guessed, reveal the answers and share your memorable moments with your team.

- Faculty lead a brief discussion: What is the difference between empathy and sympathy?

    o Empathy Short: https://www.youtube.com/watch?v=1Evwgu369Jw

Assign Midterm Essay [5 min]

- Point to stress: this concludes the sociotechnical readings portion of the class. Now we want students to go back and synthesize across multiple readings as well as their own personal experience.

Break [10 min]

Peer Infographic Feedback [30 min]

- This activity builds on the rapid prototyping gallery walk from the previous week. The goal is to have students provide feedback to each other on their draft infographics.

- Have students pull up infographic drafts on computer.

- Faculty put grading rubric on screen and remind students what is supposed to be in the infographic.

- Prompt students to swap their infographics with another team and let them grade the draft.

    - Prompt: Tell the team what's working and what isn't working [8–10 min].

    - Share findings with other team [8–10 min].

- Faculty lead a discussion on common themes that people noticed, for example missing items, things that worked well, and so forth.

Working Time and Faculty Team Meetings [30 min]

- Taking the feedback they got from their peers, student teams regroup and decide what are the next steps on their project. At the same time faculty make the rounds to chat with the teams and give input on their selected project, infographics, and answer any questions.

- Prompts for students:

    ○ Get back with your team for teamwork

    ○ What did the external evaluation find?

    ○ What changes might this suggest?

    ○ What's next step in your workplan?

    ○ What questions do you have for us?

Discipline Breakout Session [30 min]

- This is the last week of socio-technical reading and discussion. The next third of the class is focused on building a drone. We have found that this is a good time to pull students aside to talk about their experience so far. Doing so in a discipline-divided discussion facilitates frank discussions about the class, concepts, and team dynamics.

- Note: This exercise requires a second classroom.

- Faculty split and we lead an open forum where we provide the following prompts:

    ○ How are things going?

    ○ What is unexpected?

    ○ New perspectives?

    ○ Something you expected, that hasn't happened.

    ○ Something unexpected, that has.

    ○ How's your team working?

    ○ How do you talk about this class to students in our home department/school?

- Remind students of assignment due next week (midterm essay).

WEEK 6: DRONES LAB BUILD DAY 1/2

Class Prep

Assignments Due:

- Midterm essays

Before Class:

- Organize drone kits and be ready to hand them out to teams.

- Grade and return (pass/fail) previous weeks reflections.

Bring to class:

- Drone kits

By the end of this lesson students should be able to:

- Explain to a peer the fundamental engineering principles behind drones.

- Describe the major components of a drone and how they fit together.

- Craft a plan of attack for an unfamiliar project.

Class Content

Class Intro and Transition Signaling [10 min]

- Spend a few minutes at the start of class signaling to the students that (1) we appreciate their efforts so far, (2) they've done good work on getting midterm essays completed, and that (3) we are moving away from the reading heavy portion of the course.

Drone Related Engineering Concepts Presentation [60 min]

- This presentation involved a broadly accessible overview of the major principles involved in quadcopter flight. You will need to tailor this presentation to meet the backgrounds of your students—a delicate balance when there are engineers and non-engineers in the room

- We have had success taking an active learning approach and asking Engineering School students to explain certain basic topics (For example: power, current, and voltage) to their non-engineering peers. We then give a mini lecture that builds on that foundational knowledge, thereby engaging both the peace and engineering students.

- The goal is to give all students some exposure to these concepts, have a common vocabulary, and provide resources for them to learn more.

- A few topics to consider:

  ○ Lift

  ○ DC motors

  ○ Batteries

  ○ Flight control boards

  ○ Wireless communication

  ○ Quadcopter control algorithms

- Two Notes:

  ○ If you use the drone kit we've developed with NewBeeDrone, contact us for plat-form-specific lesson plans.

  ○ We have had success integrating this module with the unboxing activity so it becomes a more tangible experience for students. Consider adopting that approach.

Break [10 min]

Introduce Drone Kits [15 min]

- Here we talk about our own experience building the drones we used to get the class started. We showed a brief time-lapse video from that experience. The punchline is that we've been there.

- Before handing out kits be sure to clarify class goals for today:

  ○ Clear progress building *your own* drone.

  ○ Everyone can fully explain all drone components and how system was put together.

- Getting Started:

  ○ Engineers, restrain yourselves! Hands off other people's drones!

  ○ You should partner up with someone else! This will be easier if you are working together.

    - NOTE: We have run this class in a number of configurations, including two students working on one drone, and each student working on their own drone. In both cases we ensured that partnerships were comprised of students from different schools, and that pair worked within a table-based team context.

- Discuss Safety: For the small indoor drones we recommend using, safety is not a major concern but there are a few reminders:
    - Batteries can explode during charging or if shorted—be careful!
    - Propellers spin fast, so watch your fingers, eyes, and hair!

Unboxing, Inventorying, and Making a Plan [30 min]

- This activity is an explanation of all components and how they should be integrated—remind students not to start building yet!

- Challenge students to figure out what all the components are and think about how they might connect together. Do an inventory—name components. Look things up online. We have consistently chosen to delay providing any instructions, in an effort to push the students to do their own learning

- Ask students (in pairs) to come up with a plan of how they are going to build the drones and then present this to instructors when they are ready

Ready, Set, Build! [Rest of Class]

- Hand out additional resources to scaffold the build (custom tailor to your drone).

- Students work on building drones for the remainder of class.

- In our experience, a properly scoped build will take students a class and a half to finish. So by the end of Build Day 1, most of our students got 80–90 of the way through the physical assembly of their drone. This time will vary by build complexity.

Team Checkin (while students building)

- While the teams are building drones, faculty circle around and discuss with teams their progress on the infographics assignment, which is due in the following week.

Class Conclusion [5 min]

- How are things going in your teams?

- Drones should be 75% done by next week.

WEEK 7: DRONES LAB BUILD DAY 2/2

Class Prep

Assignments Due:

- Infographic final draft

- Readings (but no reflection)

Before Class:

- Grade and return midterm essays

Bring to class:

- Your own drone for flying as an example

By the end of this lesson students should be able to:

- Describe the final deliverables of the course.

- Program open-source software to control a quadcopter.

- Debug common challenges found when building a quadcopter.

- Fly a quadcopter.

Class Content

Final Project Introduction [20 min]

- In this session we dive into details of exactly what we are expecting student to deliver for the final project. We go through each of the final deliverables as well as all of the interim deliverables.

- We also spend time discussing the concept of the Minimum Viable Product. We show this video from Steve Blank.

- We set aside a modest budget to help students acquire resources necessary for their MVP assignment. While this will vary by context, we set an informal cap of $50 per team, but find that few students draw on this resource.

Ready, Set, Build (Continued)! [155 min]

- The remainder of class continues where students left off with the build from the previous week.

- For our builds so far this usually means 30–60 min spent assembling hardware with the remainder of the class spent in software programming the drone.

- Toward the end of class many students should have finished their drones and can start flying them.

Team Checkin (while students building)

- While the teams are building drones, faculty circle around and discuss with teams their current progress in narrowing their project focus.

Class Conclusion [5 min]

- Highlight that next week we will be talking about entrepreneurship.

- Emphasize that class time for the drone builds is completed; any additional building needed must be completed outside of class.

WEEK 8: ENTREPRENEURSHIP 101

Class Prep

Assignments Due:

- Reading and videos (no reflection)

Before Class:

- Grade and return Infographics

Bring to class:

- Nothing

By the end of this lesson students should be able to:

- Describe the major elements of social model canvas.

- Complete a social model canvas for their chosen project.

- Explain to a friend what an entrepreneurial mindset is and why it is useful.

Class Content

Infographic Feedback [45 min]

- Faculty circulated to give teams feedback on their Infographics, with particular emphasis on the fact that this effort is the first draft of the arguments they will make in the end-of-semester pitch presentations.

Social Model Canvas [120 min]

- This exercise is intended to do introduce students to key concepts in Burkette's *Social Business Model Canvas.*

  ○ This builds on the book Business Model Generation by Alex Osterwalder and Yves Pigneur. There are many different ways to teach the business model canvas, https://strategyzer.com/ has lots of good resources for those unfamiliar with this tool.

- We begin with a presentation on the Social Model Canvas.

- We used the example of M-Pesa, a successful social enterprise in Kenya, to discuss the social model canvas.

  ○ Hughes and Lonie - M-Pesa Case Study (here)

  ○ Video 1 from USAID

  ○ Video 2 from Safaricom

- We suggest you structure class by providing a brief overview of the case study the students read for class and then walking them through each segment of the business model canvas using MPesa as an example.

- After introducing each of the concepts (on a single slide with a simple example) we ask the teams to work together to address their team's response to the prompt: *What is your team's* _____ [*e.g., Key Partner, Value Proposition, etc.*]. This work can be captured on Sticky Notes, in order to emphasize the provisional stage of this exercise.

- To keep the structured time short, have the students focus on either Impact or Commerce statements as you explain the BMC. Tell them this is just to get a sense of each of these components and they will then have time at the end of class to work uninterrupted on the full canvas.

- By the end of the section teams should have a good sense of how to complete their canvas.

- We take a 10-min break halfway through this activity.

Canvas Share-Out [20 min]

- Students shared their prototype canvases with the class.

- Consider having each team focus on different segment to make sure the process is engaging.

WEEK 9: PRESENTATION PRACTICE AND FLIGHT COMPETITION

Class Prep

Assignments Due:

- Team Workplan presentation

- Expert piloting skills

Before Class:

- Setup obstacle course

Bring to class:

- Obstacle Course Materials, e.g., gaffers tape, ladder, hula hoops, twine, scissors.

- Drone (with FPV setup, if you have one).

By the end of this lesson students should be able to:

- Fly a drone through an obstacle course.

- Communicate their work plan for a complicated multi-week project.

Class Content

Team Workplan Presentations [60 min]

- These presentations are aimed at getting students to think seriously about how they will complete the final project. The public performance aspect encourages taking the assignment seriously. The presentation practice is also useful as they are coming closer to the end of the semester when they will have to deliver their final pitch.

- Presentations by each team. Each presentation is at most 10 min, roughly 7 sides per team, with a few minutes for Q+A.

Break [15 min]

- During the break faculty QUICKLY Develop Feedback for (1) the class broadly and (2) each of the teams in particular.

Feedback/Social Model Canvas Workshop [45 min]

- In this activity faculty circulate to give students feedback about their work plans. At the same time students are working on the social model canvas assignment that is due the following week.

- Before starting the feedback session, briefly recap the social model canvas assignment.

- Provide broad feedback on trends across teams [5–10 min].

- Individual team feedback [5 min per team].

Drone Competition [60 min]

- The competition is a chance to have some fun and reap the benefits of their difficult build process. Note that we have run this both indoors and outside, we vastly prefer the indoor environment as it is far more controlled.

- Before the official racing begins, give students a few minutes to prep their drones for competition—there are often a few kinks to work out.

- At the same time make any last minute adjustments to the competition course. We have constructed the course a few different ways. Indoor obstacle courses are easy to construct using found-material like desks and chairs, as well as hula hoops, which we have laid on the floor (for take-off and landing exercises) as well as hung from the ceiling. Materials you will need: String, Gaffers tape, Ladder, Hula hoops.

- While there is no wrong way to approach the competition, we've settled on the following:

  ○ Start with Time Trials—each student attempts to finish the course as quickly as they can. We are typically lenient and give students several attempts.

  ○ Final Race:

    - The final race takes place between the students with the top times in the Time Trial. How many are chosen depends on factors relevant to the instructor. In our case we found that only a handful of were able to complete the Time Trial in a reasonable amount of time, for example.

- The final race can be a slightly more difficult version of the same obstacle course deployed in the Time Trial, or it can be a significantly more difficult challenge. This will depend on student ability.

  ○ Trophy—We recommend giving the winners a toy trophy or certificate in a light-hearted nod to their achievement.

WEEK 10: PITCHING 101 + WORKING DAY

Class Prep

Assignments Due:

- Social model canvas

Before Class:

- Nothing

Bring to class:

- Enthusiasm

By the end of this lesson students should be able to:

- List at least three key elements of a successful pitch.

- Define a minimum viable product and explain how they are useful.

Class Content

Pitching 101 [30 min]

- In this activity we discuss how to give a compelling pitch. The specifics of this presentation will depend on what you are looking for in your pitches, but we highlight the following themes:

  - Presentations are stories—They should have a clear message, logical flow, and bring the audience along.
  - Slides are secondary—What's most important is the speaker. Slides should be designed last and should complement the speaker's message. Minimize text and use high quality images.
  - Consider your audience—The point of the pitch is to convince someone to do something. What is your call to action?

Working Sessions [Rest of Class]

- At this point, the rest of the semester is primarily work time for the students with individual faculty coaching for each team.

- Before turning teams loose remind them that the next deliverable is the message map—the blueprint of their pitch.

- We also emphasize that they need to be working on their MVP.

Review Social Model Canvas

- Before circulating to discuss the team's individual progress, we do a quick review of their social model canvas. Based on what we find, we have a launching off point for our discussion with each team.

WEEK 11: WORKING DAY

Class Prep

Assignments Due:

- Message Map

Before Class:

- Grade social model canvas

Bring to class:

- Excitement!

By the end of this lesson students should be able to:

- Collaborate on an interdisciplinary team to solve problems and communicate ideas in a compelling way.

Class Content

Team Working Time [5:30–6:00pm]

- Remind students that the term is almost over.

- Open discussion for questions, puzzles, and obstacles.

- Explain draft pitching practice for next week (see Week 12 for format).

- Encourage students to work on next week's deliverables:

  - Draft pitch
  - MVP 75% complete

- Faculty circulate to meet with teams one-on-one and provide additional feedback.

WEEK 12: DRAFT PITCHES

Class Prep

Assignments Due:

- Draft pitch + 75% Complete MVP

Before Class:

- Review and grade message maps.

Bring to class:

- Critical eye for providing pitch feedback.

By the end of this lesson students should be able to:

- Receive and operationalize feedback on their pitch/solution.

- Give constructive and actionable feedback to others on their pitch/solution.

Class Content

Pitch practice [Duration of Class]

- In this activity each team pitches to faculty and one other team. After their pitch they receive detailed suggestions from their peers and from faculty.

- Example Schedule:
  - Teams 6 and 1 enter the room;
  - Team 1 pitches, Team 6 listens.
  - Team 6 presents their thoughts and engages in conversation with Team 1;
  - Team 6 leaves;
  - Faculty provide additional feedback to Team 1 and make plan for next steps;
  - Team 2 enters and pitches, Team 1 listens;
  - Cycle continues through all teams until Team 6 pitches to Team 5 and receives feedback.

Working Session

- While two teams are out of the classroom for pitch practice, the remaining teams are collaborating to refine pitches and update MVP, as necessary.

- At this point we have found that there tend to be a team or two struggling with their MVP, presentation, team dynamics, or all three.

WEEK 13: WORKING DAY: PROTOTYPE FINALIZATION AND PITCH FEEDBACK

Class Prep

Assignments Due:

- Revisions as discussed after last week's draft pitch.

Before Class:

- Catch up on all that grading you have been putting off!

Bring to class:

- Some good music to help students push through end of semester crunch.

By the end of this lesson students should be able to:

- Collaborate on an interdisciplinary team to solve problems and communicate ideas in a compelling way.

Class Content

Team Working Time [Duration of class]

- Remind students that the term is almost over!

- Review logistics for final presentation—in particular remind students what external judges are coming to see them present.

- Open discussion for questions, puzzles, and obstacles.

- Meet with teams one-on-one.

WEEK 14: PITCH DRONE APPLICATION FOR SOCIAL GOOD + COURSE WRAP UP

Class Prep

Assignments Due:

- Final Pitch and MVP

Before Class:

- Double check A/V logistics

Bring to class:

- Small award for the winning team (can be funny/silly)

- Student evaluations

- Food and drink

- Judging rubrics

By the end of this lesson students should be able to:

- Present their final pitch with confidence and professionalism.

- Respond to detailed questions about their drone for good.

Class Content

- Team pitches, followed by a pizza party, are the only thing that happens on the final day of this class. We organize these presentations in the form of a pitch competition, and invite members of the faculty, staff, or local industry to serve alongside ourselves as judges.

- After each team's presentation the panel of judges probe the students with detailed questions.

- After all of the teams have presented, the judges step out to deliberate and choose a winning team. This is also a great time to have the students complete course evaluations.

- Judges reconvene and announce the winning team. Each team member of the winning team receives a small award.

- We have done our best to make this a fun conclusion to the class. We hold this final class in a formal theater on campus and open the event to the campus community.

- While we had originally hoped to conclude the semester with a debrief in the second half of this class, we have not found this to work, as the students are excited and relieved, and the guest judges are interested in hanging out. We've found pizza to be a more fitting conclusion. The students enjoy decompressing and networking with the industry judges.

APPENDIX   3

Mapping Learning Outcomes to Program Outcomes

We have students from three different degree programs in this class. Below you can see how our class learning outcomes might map onto your degree program.

1. Define sociotechnical dualism and explain to a peer why they should approach problem solving from a sociotechnical perspective:

2. Debate issues related to social change; social innovation; power, politics, and inequality; and emerging technologies;

3. Assemble, test, and fly a quadcopter using off the shelf components and open source software;

4. Explain to a friend the major ethical concerns related to drones;

5. Describe the key regulatory guidelines related to drones;

6. Design a drone that can have a positive impact on society (with a critical understanding of the ways drones can negatively impact society);

7. Collaborate on an interdisciplinary team to solve problems and communicate ideas in a compelling way; and

8. Acquire new knowledge to address problems outside of their comfort zone.

Learning Outcome	ABET	MAPJ	MASI
1 Define sociotechnical dualism and explain to a peer why they should approach problem solving from a sociotechnical perspective	2,4	4,5	2
2 Debate issues related to social change; social innovation; power, politics, and inequality; and emerging technologies	2,3,4	2,1	2
3 Assemble, test, and fly a quadcopter using off the shelf components and open source software	1	/	/
4 Explain to a friend the major ethical concerns related to drones	2,4	5	/
5 Describe the key regulatory guidelines related to drones	2	3	4
6 Design a drone that can have a positive impact on society (with a critical understanding of the ways drones can negatively impact society)	1,2,	4	3,4
7 Collaborate on an interdisciplinary team to solve problems and communicate ideas in a compelling way	1,3,5	2	1,4
8 Acquire new knowledge to address problems outside of their comfort zone	7	5	3

ABET Outcomes (BS/BA in Engineering) [133]

1. An ability to identify, formulate, and solve complex engineering problems by applying principles of engineering, science, and mathematics.

2. An ability to apply engineering design to produce solutions that meet specified needs with consideration of public health, safety, and welfare, as well as global, cultural, social, environmental, and economic factors.

3. An ability to communicate effectively with a range of audiences.

4. An ability to recognize ethical and professional responsibilities in engineering situations. and make informed judgments, which must consider the impact of engineering solutions in global, economic, environmental, and societal contexts.

5. An ability to function effectively on a team whose members together provide leadership, create a collaborative and inclusive environment, establish goals, plan tasks, and meet objective.

6. An ability to develop and conduct appropriate experimentation, analyze and interpret data, and use engineering judgment to draw conclusions.

7. An ability to acquire and apply new knowledge as needed, using appropriate learning strategies.

MAPJ Outcomes (USD Master of Arts in Peace & Justice)

1. Identify and analyze the roots and drivers of violence, oppression, and injustice together with peacebuilding strategies intended to address them.

2. Integrate knowledge and experience for problem solving.

3. Become an effective communicator with the ability to shape a compelling vision and lead effective teams.

4. Gain skills and competencies connecting individual career interests and market demands in the fields of peace, justice and social change.

5. Develop the moral imagination to address complex social problems and build positive peace.

MASI outcomes (USD Master of Arts in Social Innovation)

1. Combine leadership, peacebuilding, and business skills into a holistic approach that works toward innovative solutions for good.

2. Contextualize social issues from a variety of perspectives including cultural, political, and socio-economic variables.

3. Use the knowledge to formulate an action plan to meet emerging challenges such as refugee and migration flows, environmental degradation, and humanitarian crises.

4. Become proficient in sustainable business model design, leadership, communication, human-centered design, and problem solving.

5. Increase your potential to develop innovations that will help communities to address poverty and social injustices.

References

[1]   V. W. Setzer, "A critical view of the 'One Laptop per Child' project," in *12th World Multiconference on Systemics, Cybernetics and Informatics*, 2009, p. 6. 1

[2]   C. Baillie and G. Catalano, *Engineering and Society: Working Towards Social Justice, Part III: Windows on Society*, vol. 4. San Rafael, CA:Morgan & Claypool Publishers, 2009. DOI: 10.2200/S00195ED1V01Y200905ETS010. 2, 45

[3]   D. Riley, *Engineering and Social Justice*. San Rafael, CA: Morgan & Claypool Publishers, 2008. DOI: 10.2200/S00117ED1V01Y200805ETS007. 2, 81

[4]   J. A. Leydens and J. C. Lucena, *Engineering Justice: Transforming Engineering Education and Practice*. Hoboken, NJ: Wiley, 2017. DOI: 10.1002/9781118757369. 2, 81

[5]   G. D. Hoople, J. A. Mejia, D. A. Chen, and S. M. Lord, "Reimagining energy: deconstructing traditional engineering silos using culturally sustaining pedagogies," in *ASEE Annual Conference Proceedings*, 2018. 4

[6]   G. D. Hoople, J. A. Mejia, D. Chen, and S. M. Lord, "Reimagining energy year 1: Identifying noncanonical examples of energy in engineering," in *ASEE Annual Conference & Exposition*, 2019. 4

[7]   A. Choi-Fitzpatrick, "Drones for good: Technological innovations, social movements, and the state," *Journal of International Affairs*, 68. *Journal of International Affairs Editorial Board*, pp. 19–36, 2014. 4

[8]   A. Choi-Fitzpatrick, T. Juskauskas, and M. Sabur, "All the protestors fit to count: Using unmanned aerial vehicles to estimate protest crowd size," *Interface*, 10(1–2) pp. 297–321, 2018. 4, 5, 55, 57

[9]   A. Choi-Fitzpatrick and T. Juskauskas, "Up in the air: Applying the Jacobs Crowd Formula to drone imagery," *Procedia Eng.*, 107, pp. 273–281, 2015. DOI: 10.1016/j.proeng.2015.06.082. 4, 54, 57

[10]   A. Choi-Fitzpatrick, *The Good Drone: How Social Movements Democratize Surveillance*. Cambridge, MA: MIT Press, 2020. 4, 5, 50, 57, 87, 88

[11]   C. Roberts, M. Huang, S. Lord, R. Olson, and M. Camacho, "NSF IUSE/PFE RED: Developing Changemaking Engineers #1519453." 4

[12]   A. Choi-Fitzpatrick, "Drones for good: Technological innovation, social movements and the state," *Columbia J. Int. Aff.*, 68(1), pp. 1–18, 2014. 5, 50, 55

[13]   G. D. Hoople and A. Choi-Fitzpatrick, "Engineering empathy: A multidisciplinary approach combining engineering, peace studies, and drones," *ASEE Annu. Conf. Proc.*, 2017. 5, 45, 88

[14]   A. Choi-Fitzpatrick, E. Cychosz, M. Duffey, S. Siriphanh, H. Watanabe, J. Holland, D. Chavarria, J. P. Dingens, K. Koebel, M. Y. Tulen, T. Juskauskas, and L. Almquist, "Up in the air: A global estimate of non-military drone usage," *Good Drone Lab, Kroc Sch. Peace Stud. Cent. Media, Data, Soc. Cent. Eur. Univ.*, 2016. DOI: 10.22371/08.2016.001. 5, 57, 62

[15]   G. Hoople and A. Choi-Fitzpatrick, "Educating changemakers: Cross disciplinary collaboration between a school of engineering and a school of peace," in *IEEE Frontiers in Education Conference*, 2018. DOI: 10.1109/FIE.2018.8658611. 5, 45

[16]   E. Reddy, G. D. Hoople, and A. Choi-Fitzpatrick, "Engineering peace: Investigating multidisciplinary and interdisciplinary effects in a team-based course about drones," in *ASEE Annual Conference Proceedings*, 2018. 5, 45

[17]   L. E. Grinter, "Summary of the report on evaluation of engineering education," *J. Eng. Educ.*, 46, January, pp. 25–60, 1955. 6

[18]   D. H. Jonassen, "Engineers as problem solvers," in *Handbook of Engineering Education Research*, A. Johri and B. M. Olds, Eds. New York:Cambridge University Press, 2014, pp. 103–118. DOI: 10.1017/CBO9781139013451.009. 6

[19]   J. Trevelyan, *The Making of an Expert Engineer*. Leiden, The Netherlands: CRC Press, 2014. DOI: 10.1201/b17434. 6

[20]   V. Kaptelinin and B. A. Nardi, *Acting with Technology: Activity Theory and Interaction Design*. Cambridge, MA: MIT Press, 2006. DOI: 10.5210/fm.v12i4.1772. 7

[21]   W. E. Bijker, T. P. Hughes, and T. Pinch, *The Social Construction of Technological Systems: New Directions in the Sociology and History of Technology*. Cambridge, MA:MIT Press. 1993. 7, 52

[22]   B. Latour, *Reassembling the Social: An Introduction to Actor Network Theory*. Oxford, UK: Oxford University Press, 2007. 7

[23]   L. Winner, *The Whale and the Reactor: A Search for Limits in an Age of High Technology*, vol. 29, no. 1. Chicago, IL: University of Chicago Press, 2010. 7, 52

[24] T. E. McDonnell, *Best Laid Plans: Cultural Entropy and the Unraveling of AIDS Media Campaigns*. Chicago, IL: University of Chicago Press, 2016. DOI: 10.7208/chicago/9780226382296.001.0001. 7

[25] A. Rugarcia, R. Felder, and D. Woods, "The future of engineering education I. A vision for a new century," *Eng. Educ.*, 34(1), pp. 16–25, 2000. 8

[26] K. M. Plank, *Team Teaching: Across the Disciplines, Across the Academy*. Sterling, VA: Stylus Publishing, 2011. DOI: 10.1111/teth.12151. 14

[27] D. C. Brooks, "Space matters: The impact of formal learning environments on student learning," *Br. J. Educ. Technol.*, 42(5), pp. 719–726, 2011. DOI: 10.1111/j.1467-8535.2010.01098.x. 14

[28] W. Imms and T. Byers, "Impact of classroom design on teacher pedagogy and student engagement and performance in mathematics," *Learn. Environ. Res.*, 20,(1), pp. 139–152, 2017. DOI: 10.1007/s10984-016-9210-0. 14

[29] D. C. Brooks, "Space and consequences: The impact of different formal learning spaces on instructor and student behavior," *J. Learn. Spaces*, 1(2), pp. 1–16, 2012. http://libjournal.uncg.edu/jls/article/view/285/275.14

[30] A. L. Whiteside, D. C. Brooks, and J. D. D. Walker, "Making the case for space: Three years of empirical research on learning environments," *Educ. Q.*, 33(3), pp. 1–11, 2010. https://www.researchgate.net/publication/265965269_Making_the_Case_for_Space_Three_Years_of_Empirical_Research_on_Learning_Environments. 14

[31] S. A. Ambrose, M. W. Bridges, M. DiPietro, M. C. Lovett, M. K. Norman, and R. E. Mayer, *How Learning Works: Seven Research-Based Principles for Smart Teaching*. San Francisco, CA: Jossey-Bass, 2010. 15, 16, 18

[32] C. Meyers and T. B. Jones, *Promoting Active Learning: Strategies for the College Classroom*, 22(1). San Francisco, CA: Jossey-Bass Inc, 1993. 16

[33] P. Michael, "Does active learning work? A review of the research," *J. Eng. Educ.*, 93(July), pp. 223–231, 2004. DOI: 10.1002/j.2168-9830.2004.tb00809.x. 16

[34] S. Freeman, S. L. Eddy, M. McDonough, and M. K. Smith, N. Okoroafor, H. Jordt, and M. P. Wenderoth. "Active learning increases student performance in science, engineering, and mathematics," *Proc. Natl. Acad. Sci. U. S. A.*, 111(23), pp. 8410–8415, 2014. DOI: 10.1073/pnas.1319030111. 16

[35] J. Michael, "Where's the evidence that active learning works?," *Adv. Physiol. Educ.*, 30(4), pp. 159–167, 2006. DOI: 10.1152/advan.00053.2006. 16

[36]   J. W. Thomas, "A review of research on project-based learning," *AutoDesk*, 2000. 16

[37]   J. E. Mills, D. F. Treagust, and others, "Engineering education: Is problem-based or project-based learning the answer," *Australas. J. Eng. Educ.*, 3(2), pp. 2–16, 2003. 16

[38]   C. L. Dym, A. M. Agogino, O. Eris, D. D. Frey, and L. J. Leifer, "Engineering design thinking, teaching, and learning," *J. Eng. Educ.*, 94(1), pp. 103–120, Jan. 2005. DOI: 10.1002/j.2168-9830.2005.tb00832.x. 16

[39]   J. R. Savery, "Overview of problem-based learning: Definitions and distinctions," *Essential Readings Probl. Learn. Explor. Extend.Leg. Howard S. Barrows*, 9, pp. 5–15, 2015. DOI: 10.7771/1541-5015.1002.16

[40]   J. M. and S. B. Larmer, John., J. Larmer, J. Mergendoller, and S. Boss, *Setting the Standard for Project-Based Learning*. Alexandria, VA: ASCD, 2015. 16

[41]   J. S. Krajcik and P. C. Blumenfeld, "Setting the standard for project based learning," in *The Cambridge Handbook of the Learning Sciences*, Cambridge, UK: Cambridge University Press, 2006, pp. 317–34. DOI: 10.1017/CBO9780511816833.020. 16

[42]   P. C. Blumenfeld, E. Soloway, R. W. Marx, J. S. Krajcik, M. Guzdial, and A. Palincsar, "Motivating project-based learning: Sustaining the doing, supporting the learning," *Educ. Psychol.*, 26(3–4), pp. 369–398, 1991. DOI: 10.1080/00461520.1991.9653139. 16

[43]   B. J. S. Barron et al., "Doing with understanding: Lessons from research on problem-and project-based learning," *J. Learn. Sci.*, 7(3–4), pp. 271–311, 1998. DOI: 10.1207/s15327809jls0703&4_2. 16

[44]   R. M. Harden, "Learning outcomes and instructional objectives: Is there a difference?," *Med. Teach.*, 24(2), pp. 151–155, 2002. DOI: 10.1080/0142159022020687. 16

[45]   B. Bloom, *Taxonomy of Educational Objectives*. New York:McKay, 1956. 17

[46]   M. Sweet and L. K. Michaelsen, *Team-Based Learning in the Social Sciences and Humanities: Group Work That Works to Generate Critical Thinking and Engagement*. Sterling, VA: Stylus Publishing, 2012. 17

[47]   L. K. Michaelsen, A. B. Knight, and L. D. Fink, *Team-Based Learning: A Transformative Use of Small Groups*. Sterling, VA: Stylus Publishing, 2004. 17

[48]   P. Lencioni, *The Five Dysfunctions of a Team: A Leadership Fable*. San Francisco, CA: Jossey-Bass, 2002. 17

[49]   J. Sibley, P. Ostafichuk, B. Roberson, B. Franchini, K. Kubitzv, and L. K. Michaelsen, *Getting Started with Team-Based Learning*. Sterling, VA: Stylus Publishing, 2014. 17

[50] M. L. Loughry, M. W. Ohland, and D. J. Woehr, "Assessing teamwork skills for assurance of learning using CATME team tools," *J. Mark. Educ.*, 36(1), pp. 5–19, 2014. DOI: 10.1177/0273475313499023. 17

[51] S. V Rosser, "Group Work in Science, Engineering, and Mathematics: Consequences of Ignoring Gender and Race," *Coll. Teach.*, 46(3), pp. 82–88, 1998. DOI: 10.1080/87567559809596243. 17

[52] L. L. Baird, "Do grades and tests predict adult accomplishment?," *Res. High. Educ.*, 23,(1), pp. 3–85, 1985. DOI: 10.1007/BF00974070. 18

[53] G. E. Samson, M. E. Graue, T. Weinstein, and H. J. Walberg, "Academic and occupational performance: A quantitative synthesis," *Am. Educ. Res. J.*, 21(2), pp. 311–321, 2008. DOI: 10.3102/00028312021002311. 18

[54] P. L. Roth, C. A. BeVier, F. S. . I. Switzer, and J. S. Schippmann, "Meta-analyzing the relationship between grades and job performance," *J. Appl. Psychol.*, 81(5), pp. 548–556, 1996. DOI: 10.1037/0021-9010.81.5.548/. 18

[55] M. H. Romanowski, "Student obsession with grades and achievement," *Kappa Delta Pi Rec.*, 40(4), pp. 149–151, 2012. DOI: 10.1080/00228958.2004.10516425. 18

[56] C. Pulfrey, C. Buchs, and F. Butera, "Why grades engender performance-avoidance goals: The mediating role of autonomous motivation," *J. Educ. Psychol.*, 103(3), pp. 683–700, 2011. DOI: 10.1037/a0023911. 18

[57] A. Grant, "What straight-A students get wrong," *The New York Times*, New York, 2018. 18

[58] B. E. Walvoord, *Effective Grading: A Tool for Learning and Assessment in College*. Jossey-Bass, 2009. 19

[59] A. Darder, M. Baltodano, and R. Torres, *The Critical Pedagogy Reader*. Abingdon, Oxon: Routledge, 2009. 19

[60] J. Mejia, R. Revelo, I. Villanueva, and J. J. Mejia, "Critical theoretical frameworks in engineering education: An anti-deficit and liberative approach," *Educ. Sci.*, 8(4), p. 158, 2018. DOI: 10.3390/educsci8040158. 19

[61] J. L. Kincheloe, *Critical Pedagogy Primer*. New York:Peter Lang Inc., International Academic Publishers, 2018. DOI: 10.3726/978-1-4539-1455-7. 19

[62] J. Wink, *Critical Pedagogy: Notes from the Real World*. New York:Pearson, 2005. 19

[63] C. Seehwa, *Critical Pedagogy and Social Change: Critical Analysis on the Language of Possibility*. Boca Raton, FL: CRC Press, 2012. DOI: 10.4324/9780203829219. 19

[64] G. Ladson-billings, "But that's just good teaching ! The case for culturally relevant peda-gogy," *Theory Pract.*, 34(3), pp. 159–165, 1995. DOI: 10.1080/00405849509543675. 19

[65] P. Friere, *Pedagogy of the Oppressed*, no. 1. 1970. 19

[66] G. Ladson-Billings, "Toward a theory of culturally relevant pedagogy," *Am. Educ. Res. J.*, 32(3), pp. 465–491, 1995. DOI: 10.3102/00028312032003465. 19

[67] Federal Aviation Authority, "Educational Users." [Online]. Available: https://www.faa.gov/uas/educational_users/. 25

[68] "LibrePilot." [Online]. Available: https://www.librepilot.org/. 27

[69] "BetaFlight." [Online]. Available: https://betaflight.com/. 27

[70] J. Strobel, J. Wang, N. R. Weber, and M. Dyehouse, "The role of authenticity in de-sign-based learning environments: The case of engineering education," *Comput. Educ.*, 64, pp. 143–152, 2013. DOI: 10.1016/j.compedu.2012.11.026. 34

[71] University of San Diego, "Fowler Global Social Innovation Challenge." [Online]. Avail-able: https://www.sandiego.edu/cpc/gsic/. 34

[72] "Flipgrid." [Online]. Available: www.flipgrid.com. 36

[73] "IDEO - Design Thinking." [Online]. Available: https://designthinking.ideo.com/. 36

[74] "Stanford d.school." [Online]. Available: https://dschool.stanford.edu/. 36

[75] "Design Thinking for Educators." [Online]. Available: https://designthinkingforeduca-tors.com/. 36

[76] K. Biggelow, "Ideation Techniques: Introductory Videos." [Online]. Available: https://engineeringunleashed.com/. 36

[77] IDEO, "Design Thinking Toolkit," 2012. [Online]. Available: https://www.ideo.com/post/design-kit. 37, 44

[78] J. Gomer and J. Hille, "Essential Guide to SWOT Analysis." [Online]. Available: https://formswift.com/business-plan#swotanalysis. 38

[79] E. Ries, *The Lean Startup: How Today's Entrepreneurs Use Continuous Innovation to Create Radically Successful Businesses*. New York:Crown Press, 2011. 40

[80] E. Ries, "Lessons Learned: Minimum Viable Product: A Guide," 2009. [Online]. Avail-able: http://www.startuplessonslearned.com/2009/08/minimum-viable-product-guide.html. [Accessed: 08-May-2018]. 40

[81] C. Gallo, "How to pitch anything in 15 seconds," *Forbes*, 2012. 42

[82] KEEN, "Engineering Unleashed." [Online]. Available: https://engineeringunleashed.com/. [Accessed: 09-Nov-2018]. 43

[83] J. C. Oxley, *The Moral Dimensions of Empathy: Limits and Applications in Ethical Theory and Practice*. London, UK: Palgrave, 2011. DOI: 10.1057/9780230347809. 44

[84] J. Strobel, J. Hess, R. Pan, and C. A. Wachter Morris, "Empathy and care within engineering: qualitative perspectives from engineering faculty and practicing engineers," *Eng. Stud.*, 5(2), pp. 137–159, 2013. DOI: 10.1080/19378629.2013.814136. 44

[85] S. E. Miller and N. W. Sochacka, "A model of empathy in engineering as a core skill , practice orientation , and professional way of being," *J. Eng. Edu.* 106(1), pp. 123–148, 2017. DOI: 10.1002/jee.20159. 44

[86] B. A. Karanian, A. Parlier, V. Taajamaa, and M. Eskandari, "Provoked emotion in student stories reveal gendered perceptions of what it means to be innovative in engineering," in *Proc. of the American Society for Engineering Education*, 2019. 44

[87] E. Reddy, G. D. Hoople, and A. Choi-Fitzpatrick, "Engineering peace: Investigating multidisciplinary and interdisciplinary effects in a team-based course about drones," in *ASEE Annual Conference Proc.*, 2018. 45

[88] D. Riley, *Engineering and Social Justice*, vol. 3. San Rafael, CA: Morgan & Claypool Publishers, 2008. DOI: 10.2200/S00117ED1V01Y200805ETS007. 45

[89] K. K. Kumashiro, *Against Common Sense: Teaching and Learning Toward Social Justice*. Abingdon-on-Thames, UK: Routledge, 2004. 45

[90] M. Adams and L. A. Bell, *Teaching for Diversity and Social Justice*. Abingdon-on-Thames, UK: Routledge, 2016. DOI: 10.4324/9781315775852. 45

[91] W. Ayers, J. A. Hunt, and T. Quinn, *Teaching for Social Justice. A Democracy and Education Reader*. New York:SAGE Publications, 1998. 45

[92] J. A. Leydens and J. C. Lucena, *Engineering Justice: Transforming Engineering Education and Practice*. Wiley-IEEE Press, 2018. DOI: 10.1002/9781118757369. 45

[93] A. Choi-Fitzpatrick, "Drones, Camera Innovations and Conceptions of Human Rights," in *Visual Imagery and Human Rights Practice*, S. Ristovska and M. Price, Eds. Palgrave, 2018, pp. 35–56. DOI: 10.1007/978-3-319-75987-6_3. 51

[94] P. Hitlin, "8% of Americans say they own a drone, while more than half have seen one in operation," Pew Research Center, 2017. https://www.pewresearch.org/fact-tank/2017/12/19/8-of-americans-say-they-own-a-drone-while-more-than-half-have-seen-one-in-operation/. 54

[95]   A. Atlatszo, A. Choi-Fitzpatrick, and T. Juskauskas, *WAKE UP - Tízezrek az Erzsébet hídon az internetadó ellen*. 2014. 56

[96]   A. Atlatszo, A. Choi-Fitzpatrick, and T. Juskauskas, *Ezt láthatták a madarak az internetadó elleni tüntetésből*. Atlatszo, 2014. 56

[97]   S. Milan, *Social Movements and Their Technologies*. London, UK: Palgrave, 2013. DOI: 10.1057/9781137313546. 58

[98]   D. L. Garshelis et al., "Bears show a physiological but limited behavioral response to unmanned aerial vehicles," *Curr. Biol.*, 25(17), pp. 2278–2283, 2015. DOI: 10.1016/j.cub.2015.07.024. 65

[99]   L. F. Gonzalez, G. A. Montes, E. Puig, S. Johnson, K. Mengersen, and K. J. Gaston, "Unmanned aerial vehicles (UAVs) and artificial intelligence revolutionizing wildlife monitoring and conservation," *Sensors (Switzerland)*, 16(1), 2016. DOI: 10.3390/s16010097. 65

[100]   J. C. Hodgson and L. P. Koh, "Best practice for minimising unmanned aerial vehicle disturbance to wildlife in biological field reserach," *Curr. Biol.*, 2016. DOI: 10.1016/j.cub.2016.04.001. 65

[101]   H. Weimerskirch, A. Prudor, and Q. Schull, "Flights of drones over sub-Antarctic seabirds show species- and status-specific behavioural and physiological responses," *Polar Biol.*, 41(2), pp. 259–266, 2018. DOI: 10.1007/s00300-017-2187-z. 65

[102]   V. Dodd, "Gatwick: suspects exonerated as confusion deepens over drone attack," *Guardian*, 23-Dec-2018. 66

[103]   BBC, "Gatwick flight drone near-miss 'put 130 lives at risk,'" *BBC*, 15-Oct-2017. 66

[104]   "Gatwick drones: Military stood down after airport chaos," *BBC*, 03-Jan-2019. 66

[105]   M. Evans, J. Johnson, P. Sawer, and C. Graham, "Gatwick airport drone chaos: Man, 47, and woman, 54, arrested in Crawley - latest live news updates," *The Telegraph*, 22-Dec-2018. 66

[106]   J. Lake, "Police's own drones made Gatwick chaos worse," *The Times*, 30-Dec-2018. https://www.thetimes.co.uk/article/polices-own-drones-made-gatwick-chaosworse-n53nsrfd3. 66

[107]   G. Chamayou, *A Theory of the Drone*. New York: The New Press, 2014. 70

[108]   L. Parks and C. Kaplan, *Life in the Age of Drone Warfare*. Duke University Press, 2017. DOI: 10.1215/9780822372813. 70

[109] Whittle R, *Predator: The Secret Origins of the Drone Revolution*, New York:Macmillan, 2014. 70

[110] D. Cortright, R. Fairhurst, and K. Wall, *Drones and the Future of Armed Conflict: Ethical, Legal, and Strategic Implications*, vol. 27, no. 1. Chicago. IL: University of Chicago Press, 2015. DOI: 10.7208/chicago/9780226258195.001.0001. 71

[111] Amnesty International, "Waterboarding Is Torture: 3 Things You Need To Know," amnestyusa.org, 2019. [Online]. Available: https://www.amnestyusa.org/waterboarding-is-torture-3-things-you-need-to-know/. 71

[112] B. Anderson, "Using Chili Pepper-Armed Drones to Scare Elephants Away from Africa's Ivory Poachers," *Motherboard*, 2013. [Online]. Available: https://www.vice.com/en_us/article/3dk8n8/african-poacher-poachers-want-to-scare-animals-to-safety-with-chili-pepper-armed-drones. [Accessed: 04-Jun-2019]. 74

[113] J. King, "*Sea Shepherd Drones Will Spy on Marksmen*," Aberdeen Press and Journal, August 13, 2015. 74

[114] T. Snitch, "Satellites, mathematics and drones take down poachers in Africa," *Conversat.*, 2015. 74

[115] S. M. Lord, M. M. Camacho, R. A. Layton, R. A. Long, M. W. Ohland, and M. H. Wasburn, "Who's persisting in engineering? A comparative analysis of female and male asian, black, hispanic, native american, and white students," *J. Women Minor. Sci. Eng.*, 15(2), pp. 167–190, 2009. DOI: 10.1615/JWomenMinorScienEng.v15.i2.40. 80, 81

[116] M. Dayan, M. Ozer, and H. Almazrouei, "The role of functional and demographic diversity on new product creativity and the moderating impact of project uncertainty," *Ind. Mark. Manag.*, 61, pp. 144–154, Feb. 2017. DOI: 10.1016/j.indmarman.2016.04.016. 80

[117] G. Rulifson and A. Bielefeldt, "Understanding of social responsibility by first year engineering students: Ethical foundations and courses," in *ASEE Annual Conference Proceedings*, 2014. 81

[118] G. Rulifson and A. R. Bielefeldt, "Evolution of students' varied conceptualizations about socially responsible engineering: A four year longitudinal study," *Sci. Eng. Ethics*, 25, pp. 939–974, 2019. DOI: 10.1007/s11948-018-0042-4. 81

[119] E. A. Cech, "Culture of disengagement in engineering education?," *Sci. Technol. Human Values*, 39(1), pp. 42–72, 2014. DOI: 10.1177/0162243913504305. 81

[120] D. Baum, *Smoke and Mirrors: The War on Drugs and the Politics of Failure*, vol. 75, no. 6. Boston: Little, Brown., 1996. DOI: 10.2307/20047857. 81

[121] C. Baillie and G. Catalano, *Engineering and Society: Working Towards Social Justice, Part I: Engineering and Society*. San Rafael, CA:Morgan & Claypool Publishers, 2009. DOI: 10.2200/S00136ED1V01Y200905ETS008. 81

[122] A. Revenga and M. Dooley, "Is inequality really on the rise?," *Brookings*, 2019. [Online]. Available: https://www.brookings.edu/blog/future-development/2019/05/28/is-inequality-really-on-the-rise/#cancel. 82

[123] NCD Risk Factor Collaboration, "Trends in adult body-mass index in 200 countries from 1975 to 2014: a pooled analysis of 1698 population-based measurement studies with 19· 2 million participants," *Lancet*, 10026, 2016. DOI: 10.1016/S0140-6736(16)30054-X. 82

[124] D. M. Riley, "Mindsets in engineering," in *Engineering and Social Justice*, San Rafael, CA: Morgan & Claypool Publishers, 2008, pp. 33–47. DOI: 10.2200/S00117ED-1V01Y200805ETS007. 86

[125] J. R. Brown, "Unplugging the GPS," in *Vocation Across the Academy*, D. Cunningham, Ed. New York:Oxford University Press, 2017, pp. 204–224. DOI: 10.1093/acprof: oso/9780190607104.003.0010. 86

[126] D. Green, *How Change Happens*. New York:Oxford University Press, 2016. DOI: 10.1093/acprof:oso/9780198785392.001.0001. 87, 88

[127] D. Riley, "What do we mean by social justice?," in *Engineering and Social Justice*, San Rafael, CA: Morgan & Claypool Publishers, 2008. DOI: 10.2200/S00117ED-1V01Y200805ETS007. 87

[128] D. Nieusma and E. Blue, "Engineering and war," *Int. J. Eng. Soc. Justice, Peace*, 1(1), pp. 50–62, 2012. DOI: 10.24908/ijesjp.v1i1.3519. 88

[129] M. Ganz, T. Kay, and J. Spicer, "Social enterprise is not social change," *Stanford Soc. Innov. Rev.*, 16(2), pp. 59–60, 2018. 88

[130] I. Frazier, "The trippy, high-speed world of drone racing," *The New Yorker*, 2018. 90

[131] H. Babinsky, "How do wings work?," *Phys. Educ.*, 38(6), pp. 497–503, 2003. DOI: 10.1088/0031-9120/38/6/001. 90

[132] N. Hughes and S. Lonie, "M-PESA: Mobile money for the 'unbanked' turning cellphones into 24-hour tellers in Kenya," *Innov. Technol. Governance, Glob.*, 2(1–2), pp. 63–81, 2007. DOI: 10.1162/itgg.2007.2.1-2.63. 90

[133] ABET, "Accreditation Changes | ABET," 2017. [Online]. Available: http://www.abet.org/accreditation/accreditation-criteria/accreditation-alerts/. [Accessed: 28-Feb-2018]. 132

[134] G. D. Hoople, A. Choi-Fitzpatrick, N. Parde, D. Hoffoss, M. Mellette, R. Nishimura, and V. Gutman. (2019) "About time: Visualizing time at burning man," *The STEAM Journal* 4(1, Article 5). DOI: 10.5642/steam.20190401.05. Available at: https://scholarship.claremont.edu/steam/vol4/iss1/5. 4

About the Authors

Gordon D. Hoople is an assistant professor and a founding faculty member of the Integrated Engineering Department at the University of San Diego's Shiley-Marcos School of Engineering. His work focuses on engineering education and design. He is the principal investigator on the National Science Foundation Grant "Reimagining Energy: Exploring Inclusive Practices for Teaching Energy Concepts to Undergraduate Engineering Majors." His design work occurs at the intersection of STEM and Art (STEAM). He recently completed the sculpture Unfolding Humanity, a 12-foot tall, two-ton dodecahedron that explores the relationship between technology and humanity. Featured at Burning Man and Maker Faire, this sculpture brought together a team of over 80 faculty, students, and community members.

   Austin Choi-Fitzpatrick is an associate professor of political sociology at the Kroc School of Peace Studies at the University of San Diego, and is concurrent associate professor of social movements and human rights at the University of Nottingham's Rights Lab and School of Sociology and Social Policy. His work focuses on politics, culture, technology, and social change. His recent books include *The Good Drone* (MIT Press, 2020) and *What Slaveholders Think* (Columbia, 2017) and shorter work has appeared in *Slate*, *Al Jazeera*, the *Guardian*, *Aeon*, and *HuffPo* as well as articles in the requisite pile of academic journals.

Together, Austin and Gordon co-wrote this book as equal authors, are co-directors of The Good Drone Lab, co-teachers of the course Drones for Good, co-authors of a half-dozen articles and conference papers on sociotechnical education, and co-founders of Art Builds, a cross-disciplinary art collective.

Printed in the United States
by Baker & Taylor Publisher Services